# Poisonous Tales
## A Forensic Examination of Poisons in Fiction

# Poisonous Tales
## A Forensic Examination of Poisons in Fiction

By

**Hilary Hamnett**
*University of Lincoln, UK*
*Email: hhamnett@lincoln.ac.uk*

ROYAL SOCIETY
OF **CHEMISTRY**

Print ISBN: 978-1-83916-143-8
PDF ISBN: 978-1-83767-202-8
EPUB ISBN: 978-1-83916-481-1

A catalogue record for this book is available from the British Library

The Royal Society of Chemistry is a charity, registered in England and Wales,
Number 207890, and a company incorporated in England by Royal Charter
(Registered No. RC000524), registered office: Burlington House, Piccadilly,
London W1J 0BA, UK, Telephone: +44 (0) 20 7437 8656.

Visit our website at www.rsc.org/books

Printed in the United Kingdom by CPI Group (UK) Ltd, Croydon, CR0 4YY, UK

# Preface

Poisons have been favourite weapons of writers since Ancient times. Successful poisoners were stealthy and clever, leaving no trace of a crime and adding to the mystery of the plot. Before the 19th Century, poisons were not just a problem for the victim, but also for the authorities. With no way of testing for poisons in a dead body, it was hard to pin the crime on anyone. Many poisonings also went undetected as the symptoms were so similar to natural causes, that probing questions weren't asked. Early poisons mostly came from plants or venomous animals. This meant would-be poisoners had to conceal plant materials in food or drink, or engineer an opportunity for their victim to be bitten or stung. In the 19th Century, chemists developed new ways to purify natural sources to make potent pills and powders, and poisoning became much easier. Deadly poisons such as strychnine and arsenic could be bought from pharmacists for 'pest control' with few reliable records being kept of who was buying what. This was the heyday of poisonings, and many murder mystery writers started to incorporate poisons into their plots around this time. During the same period, chemistry was catching up with the killers, and tests for poisons at the scene of the crime and in the body were being developed.

Agatha Christie's books feature many very realistically portrayed poisonings, but this book is not about crime fiction, it is about the poisonings that happen in romances, tragedies, sci-fi

Poisonous Tales: A Forensic Examination of Poisons in Fiction
By Hilary Hamnett
© Hilary Hamnett 2023
Published by the Royal Society of Chemistry, www.rsc.org

stories, gothic novels, political dramas, spy thrillers, ancient epics and even children's tales. It is also written through the lens of a forensic toxicologist who has handled many real-life cases of poisoning, and features substances that are still seen in deaths today. This book looks at 11 case histories from fiction and examines how realistically the poisonings have been portrayed. In some cases, the writer gives us the identity of the poison (including morphine, strychnine, mandrakes, digitalis, oleander, snake venom and mercury), but in others we will need to use forensic toxicology to help us identify the chemical culprit. For the earlier works of fiction, such as Shakespeare's plays, we will look at the medical knowledge of the day, along with the symptoms, to draw our conclusions. This book also looks at the methods used to detect poisons in modern forensic toxicology laboratories (spoiler alert – you can't get away with murder), even in cases where the victim is long dead and buried.

This book is not intended as an academic textbook, but might be of interest to those studying fiction or creative writing. If you are interested in forensic science, hopefully this book will give you an insight into how real poisonings are investigated by forensic toxicologists. For more information on many of the poisons and techniques used, a Glossary is provided. If you are a budding author, I hope you can gain solid scientific inspiration for your next poisonous plot. Towards the end of each chapter, we will look at real forensic toxicology cases from around the world involving each poison. Ranging from injecting mercury as an aphrodisiac to suicide in a car with a cobra, to sports doping with strychnine, we will see that real life is sometimes even stranger than fiction.

Hilary Hamnett

# Acknowledgements

I would like to thank the library staff at the University of Lincoln for helping me research this book, my sister Gillian Hamnett of the University of Oxford for proofreading each of the chapters, and my husband Duncan Garmonsway for all his support during this long process.

Poisonous Tales: A Forensic Examination of Poisons in Fiction
By Hilary Hamnett
© Hilary Hamnett 2023
Published by the Royal Society of Chemistry, www.rsc.org

# Glossary

A word shown in **bold** within the definition indicates another entry in the Glossary under that name. Chemical structures are generally shown above the entry text.

**AAS** or **atomic absorption spectroscopy** is a technique for detecting heavy metals. A liquid sample is turned into a very fine mist and set alight with a flame. Light is then passed through the sample and any metals present will absorb light of a specific wavelength, meaning they can be identified.

**Acetylcholine** is a neurotransmitter that acts on the nervous system to modulate (control) muscle movements. Drugs that interfere with its activity are called "anticholinergics". Acetylcholine chloride is used clinically for cataract surgery.[1]

Poisonous Tales: A Forensic Examination of Poisons in Fiction
By Hilary Hamnett
© Hilary Hamnett 2023
Published by the Royal Society of Chemistry, www.rsc.org

**Aconitine** is found in fresh aconite (wolfsbane or monkshood, Figure 5.1) plant material at concentrations ranging from 0.3–2%,[2] with the highest concentrations found in the root. It is one of the three main alkaloids in the plant, alongside **mesaconitine**, and **jesaconitine**.

**2-Aminothiazoline-4-carboxylic acid** is a new marker for cyanide poisoning in post-mortem blood samples. It can be detected by **LC-MS** and is much more stable than cyanide itself.[3]

**Amphetamine** is a controlled stimulant drug (known as "speed"). It is used clinically in the UK to treat narcolepsy and ADHD.[1]

**Amygdalin** is a **cyanogenic glycoside** found in the stones or pips of plants of the *Prunus* species (see Figure 10.1 for an example). Levels in fruit are typically low: apricot (8%), plum (2.5%), bitter almond (5%) and peach (6%),[4] but when eaten it is broken down by enzymes into hydrogen cyanide (HCN). The characteristic bitter almond taste comes from benzaldehyde, which is also made when the molecule breaks down.[5] Amygdalin is sold as a nutritional supplement under the name "vitamin B17" to support the immune system.[6]

**Apoatropine** is an alkaloid that can be found in plants of the Solanaceae family, such as deadly nightshade (Figure 4.2). It is a dehydrated version of **atropine** because the two differ in their chemical structures by the presence of a water molecule.

**Aspirin** is an over-the-counter medication used to treat pain, inflammation and fever. It can also be used long term to prevent heart attack. Aspirin is made from **salicylic acid**, which was first extracted from the willow (*Salix alba*) tree (see Figure 3.6).[7]

**Atropine** is the main alkaloid found in deadly nightshade (*Atropa belladonna*) alongside **hyoscine** and **scopolamine**. 98% of the alkaloids present in the plant are atropine.[8] The leaves contain 0.2–0.5% atropine,[2,9] with the ripe berries containing around 0.6%.[2] Atropine is prescribed in the UK as eye drops and for bradycardia (slower than normal heart beat).[10]

**Brucine** is found alongside **strychnine** in the seeds of the *Strychnos nux-vomica* (Figure 8.1) plant in approximately equal amounts.[11] It is less toxic than strychnine,[12] but is structurally very similar, differing only in two methoxy groups – giving it its alternative name of "dimethoxystrychnine".

**Bufadienolide** is a **cardiac glycoside** found in some plants as well as in the skin of common toads (see Figure 6.4).

**Captopril** is a heart medication derived from the venom of the Brazilian arrowhead viper (*Bothrops jararaca*, Figure 9.6). It was used successfully to treat high blood pressure, but serious side-effects mean it is rarely prescribed now.[13]

**Cardiac glycosides** are alkaloids that affect the heart, and are found in plants such as digitalis (Figure 6.1). A "glycoside" is a chemical in which a type of sugar (shown in pink) is bound to the main molecule (black) *via* a "glycosidic" bond (green). In the case of cardiac glycosides, the molecule is similar to a steroid and contains a five-membered lactone ring (blue).[14–16]

**Cerberin** is a **cardiac glycoside** found in the kernel of the fruit of the pong–pong (*Cerbera odollam*) tree (see Figure 6.3). Ingesting half-to-a-whole kernel can be enough to cause death.[17]

**Chloroform**, also known as "trichloromethane", is a liquid that evaporates into a gas very quickly. It was used in the past as an anaesthetic for operations, but is now mainly an industrial solvent. Although fiction suggests that the merest whiff of a handkerchief soaked in chloroform would be enough to incapacitate a would-be victim, in reality 5 min or more was needed to produce anaesthesia (in a willing patient, never mind someone struggling).[18]

**Chlorpromazine** is a prescription antipsychotic medication. It can also be used to treat nausea and vomiting.[1]

**Cholestyramine**, also known as "colestyramine", is used to treat high cholesterol. It binds to bile acids (the precursors of cholesterol) that have been digested, preventing them from being re-absorbed into the body.[1] It can be used to treat **digoxin** poisoning as it binds to digoxin molecules, preventing them from being reabsorbed and speeding up their elimination from the body.[19]

**Cicutoxin** is a drug long-chain highly unsaturated alcohol found in several plant species including water hemlock (*Cicuta maculate*, Figure 3.4). The roots of the water hemlock plant contain approximately 3–5% cicutoxin.[20]

**Cocaine** is a controlled stimulant drug used recreationally for it euphoric effects. Powder cocaine (cocaine hydrochloride) is snorted or injected, whereas crack cocaine (cocaine free base) is smoked. It is often adulterated with anaesthetics such as lidocaine. It is no longer used clinically in the UK. It is found in the coca leaf (*Erythroxylum coca*) plant.

**Codeine** is an opiate found naturally in opium poppies (Figure 2.1), making up around 0.5–1% of raw opium depending on soil and climate.[2,21] When used clinically, it is usually prescribed for mild-to-moderate pain in the form of the phosphate salt, and can be given as tablets or by injection.[1] Codeine rapidly converts to **morphine** in the body.

**Conhydrine** is a **piperidine alkaloid** found in the hemlock (*Conium maculatum*, Figure 3.3) plant.[22]

**Coniine** (also known as "cicutine" or "conicine") is a **piperidine alkaloid** found in the hemlock (*Conium maculatum*) plant with the plant's fruits, stems and leaves containing 0.1–2% coniine.[23,24] Like many alkaloids, coniine can occur in two optically active forms, and it is the (*R*) form that is naturally occurring.[25] It is not used clinically in the UK.[1] Coniine was the first alkaloid structure to be fully resolved.[22]

**Cyanogenic glycosides** are alkaloids containing one or two sugars connected to a molecule containing a C≡N group *via* a glycosidic linkage.[4] They can produce hydrogen cyanide (HCN) when eaten. An example is **linamarin** (shown), which is found in plants such as raw cassava. Cassava can be eaten safely by first grinding it into flour and then soaking in water, during which the HCN evaporates.[26]

**Derivatisation** involves reacting one drug with another to make a slightly modified version, known as a "derivative". This is needed for some drugs because they are unstable at the high temperatures used in **GC-MS**. In the example shown, methamphetamine is reacted with the derivatising agent TFAA. The bulkier derivative can withstand higher temperatures.

**Diazepam** is a controlled drug, commonly known as Valium. It belongs to a family of drugs known as "benzodiazepines". Diazepam is prescribed for muscle spasms, such as sciatica, anxiety and insomnia.[1]

**Dictamnine** is an alkaloid found in the root bark of the dittany (*Dictamnus albus*) plant. It is used to treat jaundice but can cause a rash if it comes into contact with the skin.[27]

**Digitoxigenin** is found alongside other **cardiac glycosides** such as **oleandrin** in common oleander (*Nerium oleander*, Figure 7.1) plants.[28]

**Digitoxin** is found alongside other **cardiac glycosides** such as **digoxin** in digitalis (foxglove) plants. It is also converted into digoxin by the body. Digitalis plants contain up to 0.4% digitoxin.[24] Digitoxin can be used clinically to treat congestive heart failure but is not prescribed in the UK.[1]

**Digoxin** is found alongside other **cardiac glycosides** such as **digitoxin** in digitalis (foxglove) plants. Digitalis plants contain 0.5–1.5% cardiac glycosides including digoxin.[11] It is also prescribed to correct irregular heartbeat.

**Dimercaprol** is an antidote to heavy metal poisoning such as mercury. Metals binds to the S atoms and are carried out of the body through the urine.[29]

**3,5-Dimethoxyphenol** is formed from **taxicatine** when the bond connected to the sugar molecule (the glycosidic bond) is broken.[30] The part of the molecule remaining when this happens is known as an "aglycone".[31,32]

**Dimethylsulfoxide** or **DMSO** is a solvent (a liquid used to dissolve other liquids) that allows drugs to be absorbed through the skin. It is also used to treat inflammation and cystitis.[24] More recently it has been found to protect cells when they are frozen at very low temperatures in cryogenics.

**Diphenhydramine** is an antihistamine used as an over-the-counter sedative or hypnotic (sleeping tablet) under brand names such as Nytol®.

**Docetaxel** is sold under the brand name Taxotere®. It is a semi-synthetic chemotherapy agent derived from the yew tree (*Taxus baccata*, Figure 3.5). It is prescribed in the UK to treat breast cancer.[1]

**Echinacoside** is found in *Echinacea* (Figure 9.4) plants, and is an established herbal remedy sold as teabags or in tablets. It is most often used to treat cold and flu symptoms, and the roots of plants can contain 0.3–1.0% echinacoside.[33] It has also been used as a traditional remedy for snake bite envenomation.

**Eleutheroside** is an alkaloid found in the roots of *Eleutherococcus senticosus* plants. It is commonly sold in ginseng supplements.

**Enzymes** are active proteins that cells use to transform other molecules *e.g.*, to turn a parent drug into a metabolite. They are natural catalysts inside the body, but are also used in things like biological washing powder.[34]

**Epicatechin** is a flavanol (a kind of plant nutrient rather than a poison) found in many different fruits. It has anti-oxidant properties and is sold as a supplement.

**Fluorescence** happens when molecules absorb light of a certain wavelength then emit the light at a different one in the form of a glow. The wavelength of the glow is often highly specific to a drug or poison (*e.g.*, LSD) and can be measured and used to detect it.

**Galanthamine** also known as "galantamine", is found in bulbs of the snowdrop (*Galanthus nivalis*, Figure 2.5) plant but is also produced synthetically and used clinically as a prescription medication for mild-to-moderate dementia in Alzheimer's disease.[1]

**γ-Coniceine** is a **piperidine alkaloid** found in the leaves and young tissues of the hemlock plant (*Conium maculatum*).[22] Hemlock can contain up to 3.5% piperidine alkaloids including **N-methylconiine, coniine** and **γ-coniceine**.[2]

**GC-MS** or **gas chromatography-mass spectrometry** is a separation and identification technique combining gas chromatography and mass spectrometry. In the first step, a mixture is heated to a high temperature (>100 °C) and carried through a column under pressure by an unreactive gas such as helium. During this process, any molecules of interest ("analytes") separate out and exit the column at different times. The separation process is guided by the lining of the column (known as the "stationary phase"), which can be made of different types of chemicals. In the second phase, as the analytes leave the column they enter the mass spectrometer and are given an electrical charge. The charged versions (ions) are then broken down into small pieces (fragments) each of which is represented by a number (the mass-to-charge ratio or *m/z*). Each molecule breaks down in a unique way to produce a pattern or "spectrum". The analyte can then be identified by comparing the spectrum to a known sample of the drug or poison (either by running the known sample through the same GC-MS or using a library of spectra). GC-MS requires some sample preparation *e.g.*, extraction before body fluid samples such as blood or urine can be analysed. Some drugs, such as amphetamine may need to be chemically modified further (**derivatised**) to protect them from the high temperatures in the instrument.

**GC–NPD** or **gas chromatography with nitrogen–phosphorus detection** This combines the gas chromatography described above with a detector for nitrogen and phosphorous atoms. The sample is burned in a flame to produce ions, which can be measured as an electric current.

**Glycine** is an amino acid. Its role in the body is as a neurotransmitter (chemical messenger) that slows down and controls the activity of muscles. Poisons such as **strychnine** affect glycine, leading to muscles becoming overstimulated.

**Heroin** is also known as "diacetylmorphine" or "diamorphine" and is produced by acetylating the **morphine** found in raw opium. Although diamorphine is occasionally still used in emergency medicine for acute pain,[1] it is now much more likely to be seen by forensic toxicologists as a drug of abuse. The term "heroin" in the UK refers to the (usually) powdered mixture of diacetylmorphine and the various cutting agents used to improve profitability. Diacetylmorphine rapidly converts to morphine in the body. As illicit heroin usually contains some acetylated codeine, **codeine** itself is often also found in toxicology cases involving heroin users.[35]

**HPLC-DAD** or **high-performance liquid chromatography with diode array detection** is a separation and identification technique. A liquid sample (a mixture) is forced down a column under pressure by a liquid mobile phase (usually a mix of buffers and organic solvents).[36] During this process, any molecules of interest separate out and exit the column at different times. The separation process depends on what is inside the column (known as the "stationary phase"). As the individual chemicals in the mixture emerge from the column they pass into a detector. The most common type of detector used with HPLC is an ultraviolet-visible or diode array detector. These shine light of different wavelengths through each of the analytes in the mixture. Each one will absorb light at a slightly different characteristic wavelength, allowing them to be identified by comparison with a known sample or reference books. Whilst a standard ultraviolet-visible detector uses one wavelength at a time, a DAD uses multiple wavelengths.

**Hydrastine** is found in goldenseal (*Hydrastis canadensis*, Figure 8.3) and has similar effects to **strychnine**. It can poison people drinking herbal teas.

**Hyoscine**: see **scopolamine**.

**Hyoscyamine**, also known as "daturine" occurs naturally in the henbane (*Hyoscyamus niger*, Figure 3.1) plant. When attempts are made to isolate it from the plant some hyoscyamine is converted into its optical isomer (mirror image), known as **atropine**.[37] All parts of the plant contain hyoscyamine,[38] with content increasing from the root (0.08%) to the leaves (0.17%) and seeds (0.3%).[2] Clinically, hyoscyamine sulfate has been used to control gastrointestinal tract problems[11,39] but it is no longer prescribed in the UK.[1]

**Hypaconitine** is found in the aconite (wolfsbane or monkshood) plant. It is more potent than both **aconitine** and **mesaconitine**.[40]

**Immunoassay**, is based on antibody–antigen binding. In forensic toxicology, the antigen is the drug or poison being targeted during the test. In a typical immunoassay used in toxicology, a small amount of the sample is added to a well that has been coated with a specific antibody (*e.g.*, morphine antibodies for an opiates immunoassay). If there is any antigen in the sample, the antibodies sitting in the well will recognise it and bind to it. Normally, this binding cannot be seen by eye, so a colour change is used to show how much (if any) antigen was present in the sample. The amount (or intensity) of the colour can be measured and converted to a number. Somewhat counter-intuitively, the more intense the colour, the higher the number, and the lower the amount of drug or poison present in the sample. A low-intensity result indicates that one or more members of a drug 'family' are present. Immunoassays require very little sample preparation, but are classed as presumptive tests and positive results must be confirmed by a more specific technique such as **LC-MS** or **GC-MS** before they can be used in Court.

**Isotaxine B**: see **taxines**.

**Jesaconitine** is found in the aconite (wolfsbane or monkshood) plant. It is one of the three main alkaloids in the plant, alongside **aconitine**, and **mesaconitine**.

**Lactate** is created by muscles during metabolism (burning of glucose). Normally it is removed from the body, but some diseases and **strychnine** poisoning can cause it to build up, eventually lowering the pH of the blood to dangerous levels.

**Laetrile** is an alternative cancer cure. Products sold as 'laetrile' can contain **amygdalin**, mandelonitrile (upper structure) or D-mandelonitrile-β-glucuronide (lower structure). All three compounds are related by metabolism. Laetrile is a controversial drug having yet to be proved effective, and also associated with several poisonings.

**LC-MS** or **liquid chromatography-mass spectrometry** is a separation and identification technique combining HPLC and mass spectrometry. As the analytes leave the HPLC column they enter the mass spectrometer and are given an electrical charge. The ions are then broken down into fragments, each of which is represented by its mass-to-charge ratio (*m/z*). The fragments then break down further *via* a "transition" that is characteristic to that molecule. For example, **cocaine** has a characteristic transition where an ion with a number of 304 breaks down into a smaller ion of number 182, plus some very small background ions (the transition is represented as: 304 → 182).[41] These transitions are monitored and compared to the behaviour of a known sample of the drug or poison. In some testing, two mass spectrometers are used, known as tandem mass spectrometry (LC-MS/MS). Using two mass spectrometers means the small background ions can be filtered out by the first instrument, making the technique more sensitive. LC-MS requires some sample preparation *e.g.*, extraction before body fluid samples such as blood or urine can be analysed.

**Lidocaine** is a local anaesthetic and usually applied to the skin as a gel or cream. It can also be used during cardiopulmonary resuscitation (CPR), when it is injected.[1] It has a similar numbing effect to **cocaine**, so is sometimes seen as a cutting agent in street **cocaine** samples.[42]

**Linamarin**: see **cyanogenic glycosides**.

**LLE** or **liquid–liquid extraction** is a sample clean-up method that uses two immiscible layers (one watery such as urine or blood, and one of organic solvent) to draw drugs and poisons out of a biological sample. To move from the blood or urine into the solvent, the drugs must be uncharged (no +ve or −ve charges on the molecule) so often a change in pH is needed as a first step. For drugs containing nitrogen (*e.g.*, most alkaloids) the sample has to be made more alkaline. Then the organic solvent is added, and this forms a layer on top of the watery sample. Although the two layers never truly mix, by shaking, stirring or vortexing, an exchange of drugs between the two can happen. After centrifuging (spinning at high speed) the organic layer is removed then injected into the **GC-MS** or **LC-MS** instrument.

**Mesaconitine** is found in the aconite (wolfsbane or monkshood) plant. It is similar to **aconitine** in terms of toxicity.[40]

**Methamphetamine** is a controlled stimulant drug. Crystals of methamphetamine hydrochloride (known as "crystal meth") are usually inhaled. It is not used clinically in the UK, but in other countries may be used as a prescription drug to treat ADHD.

**Methaqualone** is a sedative hypnotic that was used clinically and was also a drug of abuse. It has also been found as an adulterant in **heroin**.[42] It is no longer used clinically in the UK.

**Morphine** is the main opiate found naturally in the opium poppy, making up around 10–15% of raw opium depending on soil and climate.[2,21] When used clinically, it is usually prescribed for moderate-to-severe pain and in palliative care in the form of the sulfate or hydrochloride salt,[43] and can be given as tablets, injection or an oral solution.[1]

**Naloxone** is a widely used antidote to opioid overdose. It can be given as a naloxone hydrochloride injection or in a nasal spray, and reverses the respiratory depression effects of opioids such as **heroin**.

**Neriine** (also known as "conessine") is a **cardiac glycoside** found in all parts of the oleander (*Nerium oleander*) plant.[44]

   **Nicotine** is a piperidine alkaloid found in the tobacco (*Nicotiana*) plant. It is a stimulant and highly addictive. The amount of nicotine in the leaves of the tobacco plant varies widely from 0.3–7%.[45]

   **NMDA, *N*-Methyl-D-aspartate** or ***N*-methyl-D-aspartic acid** is chemically related to amino acids and binds to NMDA receptors in nerve cells. These receptors are important for controlling memory function.

   ***N*-Methylconiine** also known as "*N*-methyl coniine" or "methylconiine", is a **piperidine alkaloid** found in the fruits of the hemlock (*Conium maculatum*) plant. Hemlock can contain up to 3.5% piperidine alkaloids including ***N*-methylconiine, coniine** and **γ-coniceine**.[2]

   **Oenanthotoxin**, also known as enanthotoxin is found in the hemlock water dropwort (*Oenanthe crocata*) plant, one of the two known as "dead men's fingers".[2] It causes severe convulsions and is related to **cicutoxin**.[46]

**Oleandrin** (upper structure), found in *Nerium oleander* plants, breaks down into **oleandrigenin** (lower structure) by losing a sugar molecule (known as "deglycosylation"). Leaves from *Nerium oleander* were shown to contain 0.018 to 0.425% of the **cardiac glycoside** oleandrin.[28] Although all parts of the plant are poisonous, the highest concentration of oleandrin is found in the root.[47] The structure is similar to naturally occurring and synthetic steroids.

**Oxalic acid** is found in rhubarb (*Rheum rhabarbarum*) plants. Although the stalks of the plant are safe to eat, the leaves can cause vomiting and convulsions.

**Physostigmine** is found in the Calabar bean (*Physostigma venenosum*) plant and is also used as a prescription medicine for glaucoma.[48] It can be used to treat poisoning by substances that interfere with the transmission of acetylcholine such as **atropine** and **scopolamine**.

**Pilocarpine** is an eye medication for glaucoma and an antidote to **atropine** poisoning.[49] It has the effect of shrinking the pupils (miosis), reducing the pressure inside the eye, and can also be used for treating a dry mouth. It is an alkaloid extracted from the *Pilocarpus* plant and mimics the actions of **acetylcholine**.[48]

**Piperidine alkaloids** are chemicals with a six-membered carbon ring where one carbon has been replaced by a nitrogen.[22] They include **nicotine**, and **coniine**.

**Polyurethanes** are plastic polymers that produce hydrogen cyanide (HCN) and carbon monoxide (CO) when burned.[50] They are commonly used to make foams, but when used in soft furnishings must be treated with a flame retardant.

**Protoanemonin** is one of the veratrum alkaloids found in the hellebore (*Veratrum*, Figure 5.2) plant. It causes vomiting, muscle twitching and convulsions. Hellebore roots contain about 0.01% veratrum alkaloids.[2]

**Protopine** is a toxic alkaloid found in bleeding heart (*Dicentra spectabilis* or *Lamprocapnos spectabilis*, Figure 7.5) plants.[51] Overall alkaloid content in the plant is <0.7%.[4] It has recently been investigated as a cholesterol-lowering drug.[52]

**Putrescine** is produced when amino acids are broken down. As this typically happens when an organism dies it is responsible for the bad smell of decomposing flesh.

**Quercetin** is a flavanoid (not an **alkaloid** because it doesn't contain nitrogen) found in many vegetables and fruits with potentially beneficial health benefits. It can be used as a folk remedy for snake bite envenomation.

**Quinidine** (upper structure) is a stereoisomer of **quinine** (lower structure), which was originally derived from the bark of the cinchona tree to treat malaria.[53] It is used clinically to restore normal heart rhythms. The alkaloid content of *Cinchona ledgeriana* is 5–8%, of which three-quarters is quinine.[27] Quinidine is present in the bark at a concentration of 0.25–3%.[24]

**Retronecine** is an alkaloid found in common comfrey (*Symphytum officinale*). Overall alkaloid content in the plant is 0.02–0.29%. The plant can be made into a herbal tea, taken for aches and pains, applied to the skin for wound healing, or stewed to use as a garden fertilizer. Ingesting the plant is now not advised, as retronecine is thought to be toxic.[4,44]

**Salicylate**, also known as "salicylic acid" is found in the bark and leaves of willow trees (*Salix alba*) and was used for aches and pains.[53] It is also a breakdown product of **aspirin** and a common ingredient in anti-acne products.

**Scopolamine**, also known as "L-hyoscine", is one of the alkaloids found in the mandrake (*Mandragora*, Figure 2.2) and jimsonweed (*Datura*, Figure 7.3) plants, making up 0.4% of the mandrake root.[54,55] It is used clinically as hyoscine butylbromide for irritable bowel syndrome, and as hyoscine hydrobromide for motion sickness.[1]

**Selegiline** (sold as Eldepryl® in the UK) is used to treat Parkinson's disease. It breaks down in the body into the controlled drugs **amphetamine** and **methamphetamine**.[56]

**Sensitisation** means that every time a dose of a drug or poison is taken, the user experiences a stronger effect. It is the opposite of **tolerance**. Sensitisation is a reaction of the immune system and in rare cases can lead to serious anaphylaxis.

**Solanine** is the major active ingredient in the woody nightshade (*Solanum dulcamara*, Figure 4.4) plant and is also found in green (unripe) potatoes.[57] It is poisonous, but victims show different symptoms to deadly nightshade poisoning with the most common being stomach pain, nausea, vomiting and diarrhoea. Plants can contain up to 6.1% solanine.[4] Solanine has a similar chemical structure to steroids and also to **cardiac glycosides** such as **digoxin** and **digitoxin.**

**SPE** or **solid-phase extraction** is a type of clean-up method that is applied to samples such as blood and urine in forensic toxicology. The liquid samples are sucked through a small cartridge containing a permeable 'bed' using a vacuum. Depending on the target drug or poison, the bed can be made of different chemicals. It initially traps the chemical of interest whilst allowing others to pass through, cleaning up the sample. Finally, a solvent is drawn through the cartridge, which carries the analyte over the bed.

**Spectrophotometry** is a technique used to detect carboxyhaemoglobin (COHb) in blood. COHb is produced when we breathe in carbon monoxide and the % saturation of the blood usually correlates with the severity of the poisoning. Spectrophotometry works by shining light of different wavelengths through the sample and monitoring how much is absorbed. The amount absorbed and the wavelengths are characteristic to COHb and can be used to create a picture or "spectrum".

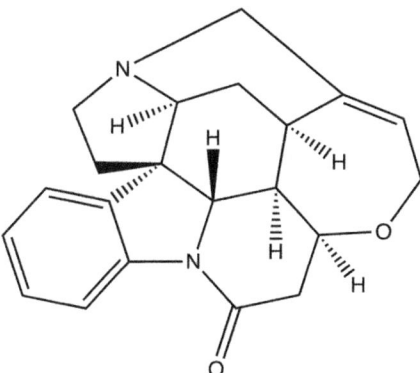

**Strychnine** is found in the seeds of the *Strychnos nux-vomica* plant (seeds contain 1–1.4% strychnine).[58] Although it was isolated from the plant in the 1800s, its chemical structure was disputed for years and not confirmed until 1952.[59]

**Taxicatine**, also known as "taxicatin" is an alkaloid found in yew (*Taxus baccata*). It is a **cardiac glycoside**.

**Taxines** are the major active compounds of yew (*Taxus baccata*) and are found at 0.1–1% in the needles. Taxine A (above) comprises approximately 1.3% of the total alkaloid content in yew.[60]

Taxine B has two isomers: taxine B (above) and isotaxine B (below).[61,62] They are the main alkaloids in yew needles, and levels in the plant vary seasonally, with the highest concentration of taxines found in January (1.2% dry weight in the needles).[30] The taxine B isomers comprise approximately 30% of the total alkaloid content.[60] It is not known which of taxine B or isotaxine B is more toxic.[61] The chemical structures of taxines are similar to the alkaloids found in digitalis, which may lead to false-positive results during **digoxin** testing (see Chapter 6).

**Thebaine** is an opiate found naturally in opium, making up around 0.2–0.5% of raw opium depending on soil and climate.[2,21] It is not used clinically in the UK.

**Thevetin A** is found in the seeds of the yellow oleander (*Thevetia peruviana*, Figure 7.2) plant. It is a **cardiac glycoside**.[4]

**TOF-MS** or **time-of-flight mass spectrometry** is a type of detector used in **LC-MS**. It works by creating ions, firing them down a long tube and timing how long they take to reach the end. Larger ions are slower than smaller ones, although even the slowest take only a fraction of a second. The time taken is characteristic to the mass of the ion and this can be compared to a library.

**Tolerance** is an effect that long-term users of certain drugs experience, whereby they need to take increasingly higher doses to achieve the same effects. These may be pleasurable or medical effects *e.g.*, pain relief. Tolerance is commonly seen for opiates and alcohol. It can occur because the body becomes more efficient at eliminating the drug from the body ("metabolic tolerance"), or because the user becomes used to the effects ("behavioural tolerance").

## REFERENCES

1. Royal Pharmaceutical Society, *British National Formulary*, BNF Publications, London, 2021.
2. D. Frohne and H. J. Pfänder, *A Colour Atlas of Poisonous Plants*, Wolfe Publishing Ltd, London, 1983.

3. J. Giebułtowicz, M. Rużycka, M. Fudalej, P. Krajewski and P. Wroczyński, *Talanta*, 2016, **150**, 586–592.
4. D. Frohne and H. J. Pfänder, *Poisonous Plants: A Handbook for Doctors, Pharmacists, Toxicologists, Biologists and Veterinarians*, Manson Publishing, London, 2nd edn, 2004.
5. J. H. Bock and D. O. Norris, in *Forensic Plant Science*, ed. J. H. Bock and D. O. Norris, Academic Press, San Diego, 2016, ch. 1, pp. 1–22.
6. M. Levine, A.-M. Ruha, K. Graeme, D. E. Brooks, J. Canning and S. C. Curry, *Chest*, 2011, **140**, 1357–1370.
7. M. J. R. Desborough and D. M. Keeling, *Br. J. Haemotol.*, 2017, **177**, 674–683.
8. F. Schneider, P. Lutun, P. Kintz, D. Astruc, F. Flesch and J.-D. Tempe, *J. Toxicol., Clin. Toxicol.*, 1996, **34**, 113–117.
9. M. Koetz, T. G. Santos, M. Rayane and A. T. Henriques, *Drug Anal. Res.*, 2017, **1**, 44–49.
10. C. M. Skinner, *Myths and Legends of Flowers, Trees, Fruits, and Plants: In All Ages and in All Climes*, J. B. Lippincott, Philadephia, 1911.
11. Medical Economics Company, *PDR for Herbal Medicines*, Medical Economics Company, Montvale, NJ, 1998.
12. F. Liu, X. Wang, X. Han, X. Tan and W. Kang, *Int. J. Biol. Macromol.*, 2015, **77**, 92–98.
13. M. Thompson, in *Pharmaceutical Journal*, Royal Pharmaceutical Society, London, 2009.
14. C. Hogue, *Chem. Eng. News*, 2005, **83**, 58.
15. R. J. Slaughter, M. G. Beasley, B. S. Lambie, G. T. Wilkins and L. J. Schep, *N. Z. Med. J.*, 2012, **125**, 87–118.
16. S. Lawrence, *Witch's Garden: Plants in Folklore, Magic and Traditional Medicine*, Welbeck, London, 2020.
17. R. G. Menezes, M. S. Usman, S. A. Hussain, M. Madadin, T. J. Siddiqi, H. Fatima, P. Ram, S. B. Pasha, S. Senthilkumaran, T. Q. Fatima and S. A. Luis, *J. Forensic Leg. Med.*, 2018, **58**, 113–116.
18. J. P. Payne, *Anaesthesia*, 1998, **53**, 685–690.
19. R. P. Henderson and C. P. Solomon, *Arch. Intern. Med.*, 1988, **148**, 745–746.
20. M. Grieve, *A Modern Herbal*, Dover Publications, New York, 2nd edn, 1971.
21. D. F. Duarte, *Rev. Bras. Anestesiol.*, 2005, **55**, 135–146.

22. H. Hotti and H. Rischer, *Molecules*, 2017, **22**, 1962.
23. J. Emsley, *More Molecules of Murder*, RSC Publishing, Cambridge, 2017.
24. *Disposition of Toxic Drugs and Chemicals in Man*, ed. R. C. Baselt, Biomedical Publications, Seal Beach, 12th edn, 2020.
25. T. Reynolds, *Phytochemistry*, 2005, **66**, 1399–1406.
26. K. Harkup, *A is for Arsenic: The Poisons of Agatha Christie*, Bloomsbury, London, 2015.
27. S. Funayama and G. A. Cordell, *Alkaloids: A Treasury of Poisons and Medicines*, Elsevier, Amsterdam, 2015.
28. Y. Gaillard and G. Pepin, *J. Chromatogr. B: Biomed. Sci. Appl.*, 1999, **733**, 181–229.
29. T. Hargreaves, *Poisons and Poisonings: Death by Stealth*, RSC Publishing, Cambridge, 2017.
30. M. Kobusiak-Prokopowicz, A. Marciniak, S. Ślusarczyk, K. Ściborski, A. Stachurska, A. Mysiak and A. Matkowski, *BMC Pharmacol. Toxicol.*, 2016, **17**, 41.
31. J. Pietsch, K. Schulz, U. Schmidt, H. Andresen, B. Schwarze and J. Dreβler, *Int. J. Leg. Med.*, 2007, **121**, 417–422.
32. F. Musshoff and B. Madea, *Int. J. Leg. Med.*, 2008, **122**, 357–358.
33. J. Liu, L. Yang, Y. Dong, B. Zhang and X. Ma, *Molecules*, 2018, **23**, 1213.
34. R. Highfield, *The Science of Harry Potter: How Magic Really Works*, Headline Book Publishing, London, 2002.
35. S. V. Konstantinova, P. T. Normann, M. Arnestad, R. Karinen, A. S. Christophersen and J. Mørland, *Forensic Sci. Int.*, 2012, **217**, 216–221.
36. J. Emsley, *Molecules of Murder: Criminal Molecules and Classic Cases*, RSC Publishing, Cambridge, 2008.
37. A. Foster, *The Medicinal Plant Collection at the University of Oxford Botanic Garden*, Wellcome Trust, Oxford, 2010.
38. A. Alizadeh, M. Moshiri, J. Alizadeh and M. Balali-Mood, *Avicenna J. Phytomed.*, 2014, **4**, 297–311.
39. J. M. Riddle, *Goddesses, Elixirs, and Witches*, Palgrave MacMillan, New York, 2010.
40. T. Y. K. Chan, *Clin. Toxicol.*, 2009, **47**, 279–285.
41. P. M. Jeanville, E. S. Estapé, S. R. Needham and M. J. Cole, *J. Am. Soc. Mass Spectrom.*, 2000, **11**, 257–263.
42. M. F. Andreasen, C. Lindholst and E. Kaa, *Open Forensic Sci. J.*, 2009, **2**, 16–20.

43. E. Hodgson, in *Progress in Molecular Biology and Translational Science*, ed. E. Hodgson, Academic Press, 2012, vol. 112, ch. 14 pp. 373–415.

44. M. R. Cooper, A. W. Johnson and E. A. Dauncey, *Poisonous Plants and Fungi: An Illustrated Guide*, The Stationery Office, Norwich, 2nd edn, 2003.

45. Z. Tassew and B. S. Chandravanshi, *Springerplus*, 2015, **4**, 649.

46. M. J. Ball, M. L. Flather and J. C. Forfar, *Postgrad. Med. J.*, 1987, **63**, 363–365.

47. P. Dey, *Biomed. Pharmacother.*, 2020, **129**, 110422.

48. W. Sneader, *Drug Discovery: A History*, John Wiley & Sons, Chichester, 2005.

49. M. R. Lee, *J. R. Coll. Physicians Edinburgh*, 2007, **37**, 77–84.

50. S. T. McKenna and T. R. Hull, *Fire Sci. Rev.*, 2016, **5**, 3.

51. M. R. Cooper and A. W. Johnson, *Poisonous Plants and Fungi in Britain*, The Stationery Office, London, 1998.

52. B. Maharjan, D. T. Payne, I. Ferrarese, M. Giovanna Lupo, L. Kumar Shrestha, J. P. Hill, K. Ariga, I. Rossi, S. Sharan Shrestha, G. Panighel, R. Lal Shrestha, S. Sut, N. Ferri and S. Dall'Acqua, *Bioorg. Chem.*, 2022, **121**, 105686.

53. The Royal Botanical Gardens Kew, *Plants+People*, Kew Publishing, London, 1998.

54. A. Chevallier, *Encyclopedia of Medicinal Plants*, DK Publishing, St Leonards, 2001.

55. M. R. Lee, *J. R. Coll. Physicians Edinburgh*, 2006, **36**, 366–373.

56. Anon, *The Lancet*, 1915, **186**, 1150–1151.

57. C. S. Hornfeldt and J. E. Collins, *J. Toxicol., Clin. Toxicol.*, 1990, **28**, 185–192.

58. I. Makarovsky, G. Markel, A. Hoffman, O. Schein, T. Brosh-Nissimov, Z. Tashma, T. Dushnitsky and A. Eisenkraft, *Isr. Med. Assoc. J.*, 2008, **10**, 142–145.

59. S. Cotton, in *Chemistry in its Element*, RSC Publishing, Cambridge, 2010.

60. C. R. Wilson, J.-M. Sauer and S. B. Hooser, *Toxicon*, 2001, **39**, 175–185.

61. L. Frommherz, P. Kintz, H. Kijewski, H. Köhler, M. Lehr, B. Brinkmann and J. Beike, *Int. J. Leg. Med.*, 2006, **120**, 346–351.

62. L. H. D. Jenniskens, E. L. M. van Rozendaal, T. A. van Beek, P. H. G. Wiegerinck and H. W. Scheeren, *J. Nat. Prod.*, 1996, **59**, 117–123.

# Contents

Poisonous Tales: A Forensic Examination of Poisons in Fiction
By Hilary Hamnett
© Hilary Hamnett 2023
Published by the Royal Society of Chemistry, www.rsc.org

# CHAPTER 1

# Introduction

If you see a term that's **bold** it's defined in the Glossary. Only the first time that the word appears in the chapter will it be indicated in this way.

## 1.1 POISONS IN FICTION

Poisons have been woven into romances, mysteries and stories for many centuries. Edgar Allan Poe is credited with writing the first fictional detective story (*The Murders in the Rue Morgue*) in 1842, which was followed quickly by series featuring Sherlock Holmes (1900) and Hercule Poirot (1920).[1] Poisonings became favourite plot devices of murder mystery writers,[2-4] with certain culprits like arsenic and **strychnine** (Chapter 8) making cameos again and again.

However, the limited scientific knowledge of some writers meant that sometimes poisons were either imaginary or not named. Even if poisons were named, the effects were often as fabulous as the story,[5] and bore little resemblance to reality. As we will see later in this book, we can guess the identity of some poisons based on the effects and what the prevailing knowledge was at the time the story was written. However, sometimes the effects are so dramatic and quick that working out the chemical

Poisonous Tales: A Forensic Examination of Poisons in Fiction
By Hilary Hamnett
© Hilary Hamnett 2023
Published by the Royal Society of Chemistry, www.rsc.org

culprit is something of a challenge. Another favourite writers' trick was to kill off characters with an unrealistic dose – the merest whiff from a chemical bottle, a pinch of powder, or single leaf of a poisonous plant (see Chapter 7). Of course, even if the writer had a good knowledge of the poison they were using, if it was left to actors to perform a play or opera they would usually go for an ostentatious death on stage (after delivering several final speeches[5]) with whatever symptoms had the most dramatic potential. We will see that some poisons act slowly and over a long period (meaning death could take months[6]) whereas others are fast-acting and really can cause death within minutes.

Then there was the hair-raising fictional identification of poisons at the scene by doctors or policemen, by sniffing, licking or rubbing suspicious samples on the lips (none of which are recommended techniques used today). Of course there are a few poisons that can be detected by smell, such as cyanide, with its characteristic bitter almond odour, but as we will see in Chapter 10, not everyone can smell it.[5] Before the 19th Century poisons were often declared 'untraceable',[4] but later detective novels introduced more science to the scene of poisonings,[7] with the development of forensic science.[8] This also coincided with more chemical poisons becoming available as household items. They started lurking in the kitchen, the medicine cabinet and the shed.[9] Before this, most poisons available to murderers or suicidal characters came from plants or animals.[10]

You might think from reading fiction that all poisonings are fatal, and indeed in Ancient times there were few or no treatments for poisoning,[9] so this *was* the usual outcome. But we now have antidotes to many poisons, some treating the symptoms and some working specifically against the chemical, such as **naloxone**, which reverses **heroin** overdose. In Chapter 6 we will meet the antidote to digitalis poisoning, and see how, somewhat unrealistically, James Bond manages to 'recover' and get back to his martini without it. Some antidotes are themselves hazardous, and doctors can end up fighting one poison with another.[11]

In fiction, the author often shows the reader the poisoner devising and enacting their sinister schemes (there is a good example of this in *The Count of Monte Cristo* in Chapter 8), but in real life, poisoners are rarely seen in action. They are usually caught through circumstantial evidence, such as buying something poisonous, or because they confess. The plots of fiction can also suggest a complex web of motives and connections between poisoner and victim, but actual motives are the same as for other killings – money, power, love, secrecy, jealousy and revenge. As in real life, poisonings are used in fiction to remove political opponents (even if they are a family member like in *Hamlet* (1600), see Chapter 3), or to dispatch an unwanted partner or a step-child who stands to inherit a fortune (see Chapter 8).[12] The calculated intelligence needed for a murder by poison makes for a good plot line,[13] but the modern-day fatal poisonings investigated by forensic toxicologists are much more likely to be accidents or suicides. Accidents are usually avoided by writers (unless as part of a vague backstory), but there are a few examples of suicides in the literature, such as the death of Shakespeare's Romeo (see Chapter 5).

In each of the chapters we will look at a different poisoning case study. Each involves a different chemical and a different manner of death (suicide, accident, murder, death in custody, *etc.*). In some case studies we know the name of the poison, such as oleander in Chapter 7, but in others we will need to use forensic toxicology to identify it. The first section of the chapter will look at the investigation of the scene, including the symptoms the victim experienced, where the poison could have come from, and who might attend the scene if it happened in modern times.

In some cases, such as *The Jew of Malta* (1589) in Chapter 2, people were declared 'dead' incorrectly. Devices that can detect faint heart signals and activity within the brain are modern inventions,[14] and when this play was written, appearing not to be breathing or looking blue were usually enough 'evidence' to pronounce someone dead. Although Coroners were established in England in 1194, not all suspicious deaths were subject to the rigorous investigation they are today. Mortality rates from disease and natural causes were so high that many deaths (particularly those of poor people) would have been accepted at face value as 'natural'. Some poisoning symptoms also looked very like diseases such as cholera.[12] Perhaps deaths of the more wealthy might have involved an "inquest", but with no way of testing for poison in the body until the late 19th Century, there was a good chance you would get away with murder.[14] Before science was routinely used, diagnosis of poisoning relied on the symptoms being witnessed and the obvious presence of poison in the stomach contents (such as plant material) so it was not particularly reliable.[15] Even today, poison rarely leaves external marks on its victims (with the exception of snake bites), but there are certain clues at the autopsy that can point to a poisoning.

## 1.2   THE POISONS BEHIND THE STORY

The second section of each chapter looks at the poisons themselves. Sometimes we know what the poison was, and in other cases we will use the symptoms to suggest the culprit. As we have seen, plants were the main source of medicines and poisons until the 19th Century. The same plant can be both medicinal (a useful herb) and toxic (a dreadful poison), it all depends on the dose. This made them good plot tools, because many of these plants, like poppies (Chapter 2), aconite (Chapter 5), henbane, hemlock and yew (Chapter 3) grow wild or were cultivated in people's gardens for 'medicinal' reasons.[12] This allowed authors to cast the net wide for the culprit.

Of course, some gardens had more poisons in them than others; as we will see in Chapter 4, religious orders such as monasteries and convents had medicinal herb or 'physic' gardens. This medical legacy was probably left by the Romans and maintained by the Christian church. In medieval times, knowledge of plants was often kept within these orders,[14] and it was common practice to give combinations of herbs in certain ways for religious purposes, and to treat those in pain.[11] Outside of the state religion there were Druids and 'wise women',[14] but during the 15–17th Centuries they were accused of being pagans or witches.[16] This risk of persecution made collecting herbal ingredients a risky business,[11] and this is where much of the folklore surrounding them has come from.[17] Instructions such as picking herbs by moonlight were probably less to do with potency (thought then to be related to the power of the moon[18]) and more to do with not being seen. Nevertheless, some knowledge was passed down between generations of women, and we see an example in fiction of a successful folk remedy using digoxin from a female healer in the 19th Century (*Silas Marner*) in Chapter 6. Other folklore sprang up to keep people (especially children) away from dangerous and valuable plants, and we see some of the weird and wonderful myths about mandrakes in Chapter 2.

Indigenous healers and herbalists also (still) prescribe herbs for various symptoms, *e.g.*, in Traditional Chinese Medicine (TCM), Kampo (from Japan) and Ayuverdic medicine (from India). A high percentage of the World's population still relies on these herbal medicines, as modern Western healthcare is too expensive, mistrusted, or not available to them. Many of the remedies still in use today can help people, but others are at best innocuous and at worst toxic,[19] as we will see in some of the modern toxicology cases we will explore later in the book. The rise of the internet has meant that herbal medicines are easier than ever to buy, and consumers are often led to believe that 'natural' is the same as 'harmless'.[20]

In the UK over time, knowledge of plants spread to the reading public as herbalists began to publish accounts of plants and their medicinal properties, known as "herbals" (the first one being published in English in the 1520s).[21] These books were designed to help people correctly identify plants and prepare and use them as medicines. Famous examples of herbal authors include John Gerard,[22] Nicholas Culpeper, Richard Banckes,[23] and William Turner.[24] Turner's was the first to be published in English (not Latin) setting a precedent for later herbalists.[9] In Turner's birthplace in Morpeth, Northumberland there is a free public medicinal garden devoted to the plants he included in his book (Figure 1.1). Some other medicinal and poison gardens have survived, such as the Chelsea Physic Garden in London, The Poison Garden in Alnwick, and The Oxford Botanic Garden and Arboretum, but these others now have entry fees.

**Figure 1.1**    (Left) The Turner Garden in Morpeth, Northumberland. Some of the images of poisonous plants in this book were taken here.

The introduction of new printing technologies in Europe saw herbals become more widely available, and by the 16th Century, pharmacies or apothecaries (see Chapter 4) began to emerge. For writers during this period such as Shakespeare, Jonson, Marlowe and Webster, herbals would have been their main source of inspiration when choosing poisons for their plots.[25,26]

The writers of herbals during the 16th Century copied extensively from earlier Roman and Greek authors such as Dioscorides, taking their knowledge at face value.[9] Unfortunately much of what was in these early herbals was incorrect and dangerous, and reputations for 'cures' probably emerged when people got better by chance after being given an ineffective herbal remedy. Placebo effects also played an important role.[27] Conversely, the dangerous effects of some toxic herbal treatments were likely missed when the patient died, apparently of their original ailment.

Herbals usually suggested ingesting medicinal and poisonous plants by making some kind of drink (tincture or decoction) out of the leaves, but some plants contain useful or deadly chemicals in their roots (*e.g.*, mandrakes in Chapter 2), seeds (*e.g.*, strychnine in Chapter 8) or fruits (*e.g.*, deadly nightshade in Chapter 4).[28] Ingestion remains the most common way to take in poisonous plants, but some can also cause skin irritation when touched, give off toxic fumes or be smoked. The chemicals in these plants are known as "alkaloids" and contain nitrogen.[29] They are designed to be a source of nitrogen for the plant (for growth and propagation[29])[30] and defend them from herbivores, so are often exceptionally bitter.[31] This means the berries of poisonous plants often have a sharp taste, leading you to spit them out,[9] although deadly nightshade berries are dangerously sweet-tasting and look a bit like blackcurrants (Figure 1.2). When used for murder, plant poisons therefore need to be disguised in strong-tasting foods such as curries.[32]

The mouth, throat and stomach are usually the first parts of the body to be affected when poisonous plants are ingested. When the rest of the digestive system comes into contact with the plant this leads to nausea, abdominal pain, vomiting and diarrhoea. This is the body attempting to expel the poison before it gets any further, but if the poison passes into the liver it is usually broken down and eliminated.[33] Some poisons produce tingling or burning sensations, or more serious spasms and convulsions

**Figure 1.2**  (Left) Deadly nightshade berries, taken at the Oxford Botanical Garden and Arboretum. (Right) Blackcurrants growing in Lincoln. Children can be poisoned by their natural attraction to berries and their lack of caution.[11]

as they target our nerves. Normally impulses travel along nerve fibres as tiny electrical currents,[34] but some poisons can intercept these causing the electrical signals to falter,[13] or as in the case of strychnine, overexcite the nerves.

Another way to kill a victim with nature was using a venomous animal, such as in *Cleopatra* (see Chapter 9). Venoms are injected directly into the blood by fangs or stings, and can affect the skin, heart and cells in ways that range from minor to fatal.[35] Unlike the alkaloids that plants use to defend themselves, venoms are used to attack and stun prey.[30,36] Some of the herbals mentioned above contained antidotes for venomous bites made from plants,[37] but as we will see in Chapter 9, these do not work and can waste valuable time before seeking effective medical treatment.

## 1.3  POISONOUS PLOTS

In the third section of each chapter we will explore other works of fiction employing the same poison; these sources can be novels, plays, films, operas or even computer games and graphic novels. Sometimes the poison is used to kill or injure a victim, in other cases it is mentioned figuratively, such as in *The Man Who Laughs* by Victor Hugo. Poisons have been popular plot devices since Ancient times and crop up in sources as broad as Greek myths,[38] the Bible, spy novels and Harry Potter.

Where we know or can guess the chemical culprit, we will examine how realistically the author has portrayed the dose, the symptoms and the way in which the poison was delivered. Most are delivered by mouth, but there are some unusual routes such as *via* the ear in *Hamlet,* or by touching, wearing or kissing poisoned objects such as robes, skulls, helmets and paintings (*e.g., The White Devil* by John Webster). Most plant poisons absorb poorly through the skin, although some can cause a local reaction (such as aconite, Chapter 5), but others like arsenic, strychnine, and several modern herbicides could cause a more extreme skin reaction or even death.[39]

### 1.4  POISONS AND THE PARANORMAL

The association between poisons and the supernatural is long, with lots of plants having supposedly magical powers, links to the Devil or the ability to conjure up creatures such as fairies (see Chapter 6).[12] Sickness itself was often put down to supernatural forces entering the body, so any cure should be likewise.[40] This was because before germs were understood, the seemingly random nature of disease could only be explained by the supernatural.[41] In this belief system, plants could protect the soul as well as the body from harm, and their divine magic required certain rituals to ensure their potency.[27]

As medical knowledge developed and the causes of sickness were found to be more earthly than divine, the magical properties of plants became old wives' tales that were passed down orally from one generation to another to keep people (especially children) away from danger.[18] Other tales come from recipes for flying ointments and magic potions that included so-called "hexing herbs" such as henbane (Chapter 3). The dual nature of many plants (healing and harming) increased the mystery around them, giving them a supposedly protective role when worn as amulets or charms.[18] Some plants such as aconite (Chapter 5) are associated with particular gods, and others bring on hallucinations, allowing the user to communicate with the divine.[42] Other myths sprung up about harvesting plants during particular phases of the moon, likely linked to the gods being absent at particular times.[18]

## 1.5 MODERN MEDICAL USES

A surprising number of deadly poisons also have medicinal uses, if the dose is right. One of the difficulties in using herbal remedies is that the potency of the plant depends on the growing conditions and the plant's age and the season. Modern medicine therefore uses the pure alkaloid, sometimes made in a lab and sometimes extracted from the plant. In these sections we will look at what good (if any) the poison can do. You may know that **morphine** from opium poppies (Chapter 2) is a common pain-reliever for example, but maybe not that the alkaloids from yew trees are useful chemotherapy agents (Chapter 3), or that chemicals from snake venoms can promote blood clotting (Chapter 9).

Plant material is still used as medicine in TCM and homeopathy, but European regulations for licensing and selling herbal products as medicines (rather than as 'supplements') have limited their widespread use.[27]

## 1.6 MODERN TOXICOLOGY CASES

In this part of each chapter we look at some of the modern cases forensic toxicologists have encountered with each poison from all over the World. As we have seen, poisonings are not always fatal, and the real-life cases we will explore range from minor symptoms to hospitalisation and death. The victims range from the very young to the very old, and the circumstances also vary, with many accidents (*e.g.*, where poisonous plants were mistaken for something edible), workplace poisonings, suicides and some homicides. The many advances in detecting drugs and poisons in the body since the 19th Century[43] mean that thankfully today's poisoners are very unlikely to get away with it. Not all 'classic' poisons are seen in modern forensic toxicology labs; arsenic, which was once a very popular poison (being easy to dissolve with no taste or smell[38]), is rarely seen nowadays and so does not feature in this book. Its decline from its heyday is probably due to tighter legal restrictions on its sale and the early development of tests for it in the body. This book also does not feature 'celebrity' or high-profile real-life poisonings. There are excellent books that already give the details of these classic cases,[44,45] whereas this book focusses on the unnoticed and perhaps less glamorous day-to-day caseload of forensic toxicologists. Spy poisonings with cyanide may feature heavily in fiction, but in real life, cyanide is most likely to be found in victims of smoke inhalation during a domestic fire (see Chapter 10).

Forensic toxicology cases follow a complex lifetime; from the scene of the poisoning, to the autopsy where the body fluid samples are taken, the initial decision about what to test for, the clean-up of what can be very dirty samples, how we identify the poison, and finally how we make sense of what the results mean. In each chapter we will explore an aspect of a forensic toxicology case, explaining the pitfalls and challenges at each point to give a sense to the reader of how these cases are handled in real life.

## 1.7  CATCHING THE CHEMICAL CULPRIT

In the penultimate section of each chapter we will look at how each specific poison might be found using the myriad ways they can now be detected in body fluids. You might be surprised to find that there is no standard set of tests for a forensic toxicology case. The decisions on what to test for are usually left up to the lab or the person submitting the samples for testing (the police or Pathologist). There are also no requirements for labs to use specific methods,[46] although some professional bodies have drawn up "best practice guidelines".[47] Not all labs can afford the top-of-the-range equipment or the staff to operate it, so this has led to a lot of variation in testing throughout the World.

Although many of the case studies from fiction in this book focus on plant poisonings, most modern cases of poisoning typically involve prescription or recreational drugs, so these will be targeted by the lab.[48] There are also differences between hospital toxicology testing and forensic toxicology testing. In hospitals, rapid drug testing of urine for a limited set of drugs (see Chapter 11) based on immunoassays (see Chapter 6) is used to make quick treatment decisions,[49] and can sometimes be falsely negative.

## 1.8 CASE CLOSED

In the final section of the chapter we will return to the case study and apply what we have learned about the poison(s) to the fictional poisoning. When passed through a modern forensic toxicology filter,[50] some are found to be unlikely or unrealistic, whereas other authors do a much better job of portraying a poisoning. Of course that doesn't mean we can't still enjoy our fiction, which is there to entertain us, but real poisoning cases often lack the finesse and romance of their fictional counterparts.[38]

## REFERENCES

1. G. S. Rao, *Int. J. Forensic Pract. Res.*, 2013, **3**, 7–12.
2. K. Harkup, *A is for Arsenic: The Poisons of Agatha Christie*, Bloomsbury, London, 2015.
3. J. F. O'Brien, *The Scientific Sherlock Holmes: Cracking the Case with Science & Forensics*, Oxford University Press, New York, 2013.
4. C. Valentine, *Murder Isn't Easy: The Forensics of Agatha Christie*, Sphere, London, 2021.
5. C. J. S. Thompson, *Poison Romance and Poison Mysteries*, George Routledge & Sons, Ltd, London, 1904.
6. A. Been, *Pharm. Hist.*, 1992, **34**, 35–39.
7. R. R. Thomas, *Detective Fiction and the Rise of Forensic Science*, Cambridge University Press, Cambridge, 1999.
8. I. D. Rae, *Chem. Br.*, 1983, **19**, 565–569.
9. M. Brown, *Death in the Garden: Poisonous Plants & their Use throughout History*, Pen & Sword Books Ltd, Barnsley, 2018.
10. B. Hubbard, *Poison: The History of Potions, Powders and Murderous Practitioners*, Welbeck, London, 2020.
11. R. Bevan-Jones, *Poisonous Plants: A Cultural and Social History*, Windgather Press, Oxford, 2009.
12. F. Inkwright, *Botanical Curses and Poisons*, Liminal 11, London, 2021.
13. D. Blum, *The Poisoner's Handbook: Murder and the Birth of Forensic Medicine in Jazz Age New York*, Penguin Books, London, 2010.
14. K. Harkup, *Death by Shakespeare: Snakebites, Stabbings and Broken Hearts*, Bloomsbury, London, 2020.

15. K. D. Watson, *Acad. Forensic Pathol.*, 2020, **10**, 35–46.
16. N. Bailey, *Chelsea Physic Garden: Connecting People with Plants since 1673*, Chelsea Physic Garden, London, 2015.
17. B. Mkhitaryan, G. Mazmanyana, H. Ghazaryan, M. Grigoryan, H. Apresyan, V. Asoyan and M. Grigoryan, *40th International Congress of the European Association of Poisons Centres and Clinical Toxicologists (EAPCCT)*, Tallinn, Estonia, 2020, p. 545.
18. F. Inkwright, *Folk Magic and Healing: An Unusual History of Everyday Plants*, Liminal 11, London, 2019.
19. J. Katz, K. Prescott and A. D. Woolf, *Am. J. Emerg. Med.*, 1996, **14**, 475–477.
20. E. Ernst, *Pharmacoepidemiol. Drug Saf.*, 2004, **13**, 767–771.
21. *The Cambridge History of Medicine*, ed. R. Porter, Cambridge University Press, Cambridge, 2006.
22. *Gerard's Herbal*, ed. M. Woodward, Senate, London, 1994.
23. Anonymous, *Herbal*, Rycharde Banckes, London, 1525.
24. W. Turner, *A New Herball*, Cambridge University Press, Cambridge, 1989.
25. M. Willes, *A Shakespearan Botanical*, Bodleian Library, Oxford, 2020.
26. A. S. Harper-Leatherman and J. R. Miecznikowski, *J. Chem. Educ.*, 2012, **89**, 629–635.
27. R. Richardson, *Britain's Wild Flowers: A Treasury of Traditions, Superstitions, Remedies and Literature*, The National Trust, London, 2017.
28. Chelsea Physic Garden, 2021.
29. Medicines from Plants, https://s3-eu-west-1.amazonaws.com/assets.botanic.cam.ac.uk/wp-content/uploads/2019/12/medicines_trail_web.pdf, accessed July 2021.
30. R. Highfield, *The Science of Harry Potter: How Magic Really Works*, Headline Book Publishing, London, 2002.
31. A. Bassolii, G. Borgonovo and G. Busnelli, in *Modern Alkaloids*, ed. E. Fattorusso and O. Taglialatela-Scafati, Wiley, New York, 2007, ch. 3, pp. 53–72.
32. T. Moffatt, *TIAFT Bull.*, 2018, **48**, 4–10.
33. M. R. Cooper, A. W. Johnson and E. A. Dauncey, *Poisonous Plants and Fungi: An Illustrated Guide*, The Stationery Office, Norwich, 2nd edn, 2003.
34. J. Howgego, in *Chemistry in its Element*, RSC Publishing, Cambridge, 2012.

35. R. M. Renneboog, in *Principles of Chemistry*, ed. D. R. Franceschetti, Salem Press, Ipswich, MA, 2016, ch. 123, pp. 383–386.
36. *Snake Venoms and Envenomation: Modern Trends and Future Prospects*, ed. Y. Utkin and A. V. Krivoshein, Nova Publishers, New York, 2016.
37. P. Dioscorides, *The Greek Herbal of Dioscorides*, Haner Publishing, New York, 1959.
38. S. Smith, *Med.-Leg. J.*, 1952, **20**, 153–167.
39. K. F. Faldetta and S. A. Norton, *JAMA Dermatol.*, 2016, **152**, 797.
40. A. C. Kail, *Med. J. Aust.*, 1983, 2, 515–519.
41. M. Chamberlin, *Old Wives' Tales: The History of Remedies, Charms and Spells*, The History Press, Cheltenham, 2020.
42. S. Schultes, A. Hofmann and C. Rätsch, *Plants of the Gods*, Healing Arts Press, Rochester, VT, 2001.
43. *Clarke's Analysis of Drugs and Poisons*, ed. A. C. Moffatt, D. Osselton and B. Widdop, Pharmaceutical Press, London, 4th edn, 2011.
44. J. Emsley, *Molecules of Murder: Criminal Molecules and Classic Cases*, RSC Publishing, Cambridge, 2008.
45. J. Emsley, *More Molecules of Murder*, RSC Publishing, Cambridge, 2017.
46. Q. Liu, L. Zhuo, L. Liu, S. Zhu, A. Sunnassee, M. Liang, L. Zhou and Y. Liu, *Forensic Sci. Int.*, 2011, **212**, e5–e9.
47. S. P. Elliott, D. W. S. Stephen and S. Paterson, *Sci. Justice*, 2018, **58**, 335–345.
48. S. P. Elliott, *Sci. Justice*, 2002, **42**, 111–115.
49. M. Kobusiak-Prokopowicz, A. Marciniak, S. Ślusarczyk, K. Ściborski, A. Stachurska, A. Mysiak and A. Matkowski, *BMC Pharmacol. Toxicol.*, 2016, **17**, 41.
50. R. Girling, in *The Sunday Times*, London, 2004.

CHAPTER 2

# The Cry of Death

If you see a term that's **bold** it's defined in the Glossary. Only the first time that the word appears in the chapter will it be indicated in this way.

> **Case History: Death in Custody**
>
> *A man is arrested and detained for multiple murders, including his own daughter and a convent of nuns. These murders are thought to have been poisonings. He is later found lifeless by his prison guard, who, presuming him dead, removes his body and dumps it outside the city walls. Some time later the man recovers consciousness and testifies as to how he faked his own death*
>
> BARABAS:    ...I dranke of Poppy and cold mandrake juyce;
> And being asleepe, belike they thought me dead,
> And threw me o're the wals:
>
> Act V, scene i
>
> *The Jew of Malta* by Christopher Marlowe,[1] 1589–90

## 2.1 THE INVESTIGATION

At first glance, it might seem implausible that a concoction of two plants could give a person a convincing appearance of death. Barabas' symptoms are not described, he is just carried

Poisonous Tales: A Forensic Examination of Poisons in Fiction
By Hilary Hamnett
© Hilary Hamnett 2023
Published by the Royal Society of Chemistry, www.rsc.org

off stage "as dead", but we can assume that they included lack of consciousness, blue skin and floppy muscles. In the absence of modern first aid training, this would have been enough to pronounce him life extinct (often referred to in death investigation paperwork as the deceased being "PLE'd"). As we will see, both of the plant ingredients in his "juyce" (mandrakes and poppies) contain alkaloids (see Chapter 1), which have the potential to cause the kind of deep sleep we would now recognise as a coma. We also know from John Gerard, a herbalist and author of *The Herball or Generall Historie of Plantes*, first published in 1597,[2] that the sleep-inducing and potentially fatal properties of these plants were scientific knowledge in Marlowe's day.

These days, deaths in custody such as a prison, youth detention centre, high-security psychiatric hospital or police cell are routinely referred to the Coroner or Medical Examiner, and may also be subject to an independent police investigation. The forensic toxicologist would be asked to determine if the deceased had used or ingested any substances that could have played a role in their death.

## 2.2 THE PLANTS BEHIND THE STORY

### 2.2.1 Poppy

Barabas' "Poppy" probably refers to the opium poppy (*Papaver somniferum*), as although there are around 70 species of poppy,[3] only one has sleep-inducing (known as "narcotic" or "hypnotic") properties. These distinctive plants are not the common red–orange field, Flanders or corn poppies (*Papaver rhoeas*) we use for remembrance (Figure 2.1, left);[4] opium poppies are more likely to have purple or white flowers (Figure 2.1, middle),[5,6] but they do give us the black seeds for our poppy seed bread and muffins (so are known as the "breadseed poppy"[6]). Although we can be very specific now about poisonous plants, in earlier times "poppy" could be used in literature figuratively to mean anything narcotic.[7] Both types of poppy grow wild in the UK on roadsides and areas of wasteland.[8]

Barabas would have got nowhere eating the petals, seeds or leaves of the poppy, as it was the opium found in the sap of seed head or capsule (Figure 2.1, right) that he needed to appear lifeless. After the petals have fallen off the plant (in July and August[2]),

**Figure 2.1** (Left) Common or field poppies flowering. (Middle) Opium poppy in flower and (right) their seed heads, taken in Lincoln. In hotter countries, the seed heads of the opium poppies grow much larger.[21]

the seed capsule is cut, and the next day the white juice (sap) or "latex" that oozes out is collected.[9] Traditionally it is dried in the sun to form a brown solid (raw opium) that is ground into a powder.[10,11] Opium contains over 40 different natural chemicals[12] (called "opiates"), the most important of which is **morphine**, named after Morpheus the Greek god of dreams (one of the effects of morphine is described as a "dreamy euphoria") who is said to have discovered the sleep-inducing properties of these plants.[13] Morphine acts on areas of the body called "receptors" (welcome sites[14]), which for opiates are found in the brain and spinal cord. If too much morphine is taken, the part of the brain responsible for regulating breathing is affected, causing it to slow down or even stop. At the same time, the person's pulse becomes weak[15] and the lack of oxygen in the blood causes the skin to turn blue (known as "cyanosis").[5,16]

Even though scientists didn't work out which chemical was responsible for opium's actions until the 1800s, its sleep-inducing and pain-relieving properties had been used since Roman times,[17] and it was known that opium could be dissolved in wine to make a drink[18] that according to Gerard can "cause sleep".[19] Paracelsus (often called "the father of toxicology") is given the credit for first dissolving opium in alcohol in 1527[20] to make

**Figure 2.2**    A mandrake plant in flower. © Chris Pole/123RF.COM.

what we now know as *Laudanum*,[†] bottles of which are on display in the Science Museum in London. Laudanum was used mainly to suppress coughs and treat pain, but opium also had other folk medicine uses, such as for cholera, gout, flu and piles.[25]

### 2.2.2   Mandrake

The other ingredient of the juice in this case was mandrake (the *Mandragora* species rather than *Podophyllum peltatum*[26]). These plants have floppy leaves, white or purple flowers in a flat rosette (Figure 2.2), yellow berries, and most importantly Y-shaped roots, that vaguely resemble the lower half of the human body.[27] We shall see how this feature gave rise to many mandrake myths and legends later on, but the name "mandragora" is said to be derived from "man dragons", mischievous creatures who could be summoned by a sorcerer.[28]

Although mandrakes have now largely fallen out of medical use, they were once highly prized for their therapeutic *and* poisonous properties. The fruit was eaten, the leaves chewed or juiced, and the bark dissolved in liquids (such as wine) to make ointments and tonics. Left to ferment, the roots could also be used to make a poisonous or sleep-inducing brew.

---

[†]Laudanum is a mixture of alcohol (usually brandy[21]), distilled water and opium[22,23] (containing about 10% opium by weight). It was widely used as a home remedy or prescribed medication in the 19th Century for diarrhoea (or as Turner puts it, "running of the belly"[24]), pain and to produce sleep, before being recognised as highly addictive and controlled in the 1970s.

Mandrake roots contain as many as 80 chemicals,[29] the most useful of which is **scopolamine** (also known as **hyoscine**). It was the late 1880s before these chemicals were isolated from the plant, and initially they were lumped together in something called "mandragorine".[30] The chemical structure of scopolamine is similar to that of **cocaine**, and it shares some of the same effects, such as delusions. In high doses, scopolamine can also cause drowsiness and even coma, but it acts in a different way to morphine on the body. Scopolamine prevents communication between nerves by blocking the action of something called a "neurotransmitter" (a small chemical messenger). There are several different neurotransmitters, but the one scopolamine acts on is **acetylcholine**, which plays a major role in attention. By interfering with acetylcholine, scopolamine can affect our memory, ability to concentrate, and alertness. It can have useful therapeutic effects however; at low doses it blocks the signals between the inner ear and the brain, preventing motion sickness.

Scopolamine is also found in two members of the nightshade family of plants. The first is *Brugmansia* or Angel's trumpet (Figure 2.3) and the second is Devil's trumpet (*Datura stramonium*), which we will meet again in Chapter 7.

**Figure 2.3**    Brugmansia plants with (left) white flowers, taken at The Poison Garden in Alnwick and (right) pink flowers, taken at Kew Gardens. They are ornamental flowers and also used in medicine.[30]

## 2.3 POISONOUS PLOTS

### 2.3.1 Dead to the World

Combining morphine and scopolamine would lead to enhanced sedation as both chemicals act together on the same parts of the brain. This would have been known in the 16th Century, as the Romans had reported that the anaesthetic power of mandrake was increased when combined with extracts from the opium poppy and henbane and hemlock, which we will meet in later chapters.[17] A sponge was soaked in the plant juices, dried and the vapour inhaled by the patient.[31] When mandrake wine was used during Roman crucifixions it was known as *morion*, or "death wine" because of its ability to make living victims appear dead.[17]

Drug-induced sleep might seem like a corny plot device, but even our greatest poets weren't above using it for its dramatic potential. Chaucer has a character break out of prison, this time by drugging his guard's wine, in *A Knight's Tale*

Soon after midnight Palamon, by aid
Of a kind friend, his way from prison made
And flies the city fast as he may go:
For he had drug'd the tippling jailer so
With a deep cup of spic'd and honied wine
And Theban opium, and narcoticks fine,

Act I, scene v
*A Knight's Tale* by Geoffrey Chaucer,[32] 1392

As we saw in Chapter 1, alkaloids are often very bitter,[33] so the honey and spices were likely added to the wine to disguise the taste of the opium. There's also a clue in *A Knight's Tale* as to where poppies may have grown in earlier times. Although they are now grown in Turkey, Afghanistan, Myanmar and Tasmania, "Theban opium" refers to an Egyptian variety of poppy grown in the Ancient city of Thebes.[34] This name was later given to another opiate, **thebaine**, which is also found in opium.[5] Alcohol in the wine also has a CNS depressant effect.

Mandrake potions were also thought to have useful medicinal effects, and one such potion is used in Shakespeare's *Cymbeline* (1610)[35] by Cymbeline's daughter Imogen. Again, mandrakes are not mentioned by name, but modern sources[36] have concluded that they were probably the source of the potion's death-like sleep effect. Imogen takes the potion believing it to be a tonic for sickness, a restorative. And in fact, from Roman times onwards, the root and bark were used in small doses as an anaesthetic, painkiller, sedative, and as a pre-medication for surgery (patients chewed on a piece of root).[17,37,38] These medications often contained multiple ingredients, including opium, henbane and hemlock, so it is difficult to pin-point the contribution of mandrake to such a concoction.

Another of Shakespeare's young female characters, Juliet in *Romeo & Juliet* (1594)[39] takes an apparently deadly potion made of un-named ingredients, which could have included mandrakes,[40] but also deadly nightshade,[41] which we will meet in a later chapter.

This potion ingredients theme also appears in *Harry Potter and the Goblet of Fire* (2000),[42] and in *The Chamber of Secrets* (1998) where mandrakes are described as "an essential part of most antidotes",[43] and later used to cure the petrified victims of a basilisk. Gerard tells us that wine made from mandrake root can be used for pain.[2]

There are also symbolic references in literature to the sleep-inducing properties of mandrakes. In *Antony & Cleopatra* (1606–7), Cleopatra asks her attendant Charmian for something to block out the time during the absence of Antony, her lover, who has been called from Egypt back to Rome by Octavian to fight Pompey's rebellion

| CLEOPATRA: | Give me to drink mandragora. |
| CHARMIAN: | Why, madam? |
| CLEOPATRA: | That I might sleep out this great gap of time, My Antony is away. |

Act I, scene v
*Antony & Cleopatra* by William Shakespeare,[44] 1606–7

It's not clear how the mandrake would have been taken by Cleopatra, but one method was to mix powdered bark with egg whites and paste it on the forehead.[6] Gerard suggests the simpler solution of just smelling the fruit to bring on sleep.[2] Cleopatra eventually dies of a snake bite, and we will learn more about how in Chapter 9.

Shakespeare refers to the sleep-inducing properties of mandrakes and poppies[45] in *Othello* (1604): Iago has persuaded Othello that his beloved wife Desdemona is being unfaithful to him, and refers to Othello's sleep being troubled with poisoned thoughts when he says

IAGO:     Not poppy, nor mandragora,
          Nor all the drowsy syrups of the world,
          Shall ever medicine thee to that sweet sleep
          Which thou owedst yesterday.

Act III, scene iii
*The Tragedy of Othello, Moor of Venice* by William Shakespeare,[46] 1604

Much further back than Shakespeare, Apuleius used mandragora in his story *The Golden Ass*, where it is made into a sleeping potion. It is supplied by a doctor as a 'poison' to a man expected to commit murder with it[47]

When this rascal was so eager to buy a deadly poison,
I thought it improper for one of my profession to provide anybody with the means of death....
So I gave him his "poison"; but it was a soporific draught of mandragora,
a proven narcotic, as you know, which induces a sleep indistinguishable from death.

10.11
*The Golden Ass or Metamorphoses*
by Apuleius (trans. E. J. Kenney),[48] Born 125 CE

## 2.3.2   The Cry of Death

As well as their sleep-inducing ways, mandrakes have a long history of superstition and folklore attached to them, and they feature in literature from classical times to the modern day. Mandrakes are not native to the Northern European climate, they grow in the Mediterranean particularly in dry river beds,[9] so as well as being useful, the cost of importing them meant they were very rare and extraordinarily expensive. Indeed, in Victorian times, mandrake was said to mean "rarity".[49]

The main legend, thought to come from the Ancient Greeks,[50] was that the mandrake (the root of which resembled a man), on being uprooted, produced a scream so powerful that it would either kill or send mad the harvester. To avoid this, you had to engage in elaborate rituals involving drawing circles, crucifixes, dogs, facing West, sprinkling bodily fluids, ivory (said to protect you from evil) and singing and dancing at the full moon with the wind in a certain direction.[30,51–53]

The cry of the mandrake is also described in the Harry Potter novel *The Chamber of Secrets*[43] and its deadly effects put to good use in the final siege of Hogwarts in *The Deathly Hallows* (2007).[54]

In *Romeo & Juliet*, the heroine, who has been ordered by her parents to marry Paris, hatches a plan to appear dead and be taken to the family tomb. On the eve of her wedding, Juliet speculates anxiously that the spirits of her ancestors may rise up

JULIET:    ...And shriek like mandrakes torn out the earth,
That living mortals, hearing them, run mad—

Act IV, scene iii
*The Tragedy of Romeo & Juliet* by William Shakespeare,[39] 1594

The noise was sometimes called a 'groan' instead, and in *Henry VI, Part 2* (1590),[55] The Earl of Suffolk says: "Could curses kill as doth the mandrake's groan".

Those in modern times who have been brave enough to try to uproot a mandrake have found that they do emit a little squeak when pulled out of the ground, like many similar plants.[6] But mandrakes also had a mystical aura[47] and were thought to shine at night like a lamp, making them easier to find in the dark.[38] This glowing in the dark led to the plant being called "Devil's candle"[6] and was used in the romance *Lalla Rookh* written by the poet Thomas Moore in 1817

Such rank and deadly lustre dwells,
As in those hellish fires that light
The mandrake's charnel leaves at night!

*Lalla Rookh* by Thomas Moore,[56] 1817

### 2.3.3 Motherhood

Mandrakes were also strongly associated with female fertility, partly because of the vaguely 'baby' shape of their roots. The root of the mandrake was coveted as a kind of charm or talisman that was placed on the mantelpiece or carried around wrapped in cotton.[6] This was because, as well as their therapeutic properties, mandrakes were strongly associated with female fertility. The herbalist Gerard, who also set out to bust many of the mandrake myths, suggests that their resemblance to humans was nothing more than a ruse perpetuated by con-artists who carved the roots into a human form. They were then sold to women looking for a fertility charm to hang over their headboard or place under their pillow. Turner also tells us that mandrakes can stop women menstruating,[24] a myth that dates back to a much earlier herbal by Dioscorides, the oldest copy of which dates from AD 512, but which was not translated into English until 1655.[57]

This myth also appears in the Bible in the story of the two wives of Jacob, Rachel (who is childless) and Leah

In the days of the wheat harvest Reuben went and found mandrakes in the field, and brought them to his mother Leah. Then Rachel said to Leah, "Please give me some of your son's mandrakes."

Genesis 30: 14
*The New Revised Standard* Bible[58]

Rachel wanted the mandrakes because the inability to conceive was considered a great misfortune, disgrace and even a curse on women at the time.[59] Leah eventually agrees, the *duda'im* or "love apples"[60] work their 'magic' and Rachel conceives two sons.[59] We are not told how the mandrakes were used, whether they were made into talisman or charm, or if a small amount of the root or apple-like fruit[5] was consumed (although apparently it had a foul stench so this may have been unlikely[52]). There is no scientific evidence for mandrakes curing female sterility, but small doses of scopolamine can lead to feelings of elation and a lack of inhibition – in other words aphrodisiac effects.

However these may not have been mandrakes at all. The original Hebrew word was translated as mandrake, but the "fragrance" of the love apple is mentioned in another part of the Bible, the Song of Solomon, an Old Testament exchange of songs between a bride and bridegroom and as Gerard points out, mandrake flowers have no noticeable aroma.

The mandrakes give forth fragrance,
and over our doors are all choice fruits,
new as well as old

Song of Solomon 7: 13
*The New Revised Standard Bible*[58]

The fertility myth also appears in Shakespeare's *Henry IV, Part 2* (1600)[61] to refer to Justice Shallow's apparently high sex drive, and is centre stage in Machiavelli's play *La Mandragola* (1524), which features a fertility-giving potion of mandrake roots. The potion is prescribed by Callimaco, who is posing as a doctor. If the beautiful Lucrezia drinks the potion, it will help her to conceive a child with her husband Nicia, with the slight snag that the first man who shares her bed after she takes the drink will die within a week[62]

CALLIMACO:     You have to understand this: that there is nothing more certain to impregnate a woman than to give her a potion made from the mandrake

Act II, scene vi
*La Mandragola* by Niccolò Machiavelli (trans. J. Crawford),[63] 1524

If mandrake roots were consumed to aid fertility, the dose would have to be right. Too much could lead to death-like sleep, or as we will see, to memory loss and even 'madness'.[25]

### 2.3.4  Madness

Scopolamine passes easily into the brain and can produce a state of mind where the user starts to hallucinate and can no longer distinguish between fantasy and reality.[64] This has given mandrake roots a reputation for causing madness, which goes back to Ancient Greece. This is probably the origin of association of mandrakes with the Devil, as madness in the Christian tradition was often ascribed to possession by evil spirits.[30]

Although the plant is not named, the "insane root" in *Macbeth* (1605) could have been mandrake, or possibly henbane,[37,38,65] which we will meet in another chapter. Early on in the play, having just encountered the three witches emerging from the mist to deliver their prediction that Macbeth will be King of Scotland, Banquo fearing that he and Macbeth may be hallucinating, says

BANQUO:    Were such things here as we do speak about?
Or have we eaten on the insane root
That takes the reason prisoner?

Act I, scene iii
*The Tragedy of Macbeth* by William Shakespeare,[66] 1605

Lady Macbeth may also have drugged Duncan's grooms with a potion of poppy and spiced milk curdled with ale or wine.[65]

Mandrakes were also associated with madness by John Webster, in another of his plays, *The Duchess of Malfi* (1612–13). The recently widowed Duchess re-marries, this time choosing a servant. Her brothers do not approve of the match and when they discover her marriage, they are enraged saying

FERDINAND:    I have this night dig'd up a man-drake
CARDINAL:    Say you?
FERDINAND:    And I am growne mad with't.

Act I, scene v
*The Duchess of Malfi* by John Webster,[67] 1612–13

### 2.3.5 Memory Loss

In the final of scopolamine's effects we will consider here, the short-term memory is disabled, leading to amnesia. We still do not completely understand the mechanism for memory loss, but it may involve the **N-methyl-D-aspartate** (**NMDA**) receptor.[68] This brings us to one of the first possible uses of mandrakes in literature: in Homer's *Odyssey*. Another name for mandrake is *Circaeon* or "Circe's plant", and in the poem, Circe (a witch) (Figure 2.4) poisons a band of Odysseus' comrades[59]

**Figure 2.4**    Circe offering the cup to Odysseus, by John William Waterhouse.

Within the forest glades they found the house of Circe,
built of polished stone...
She brought them in and made them all
sit on chairs and seats and made for them a potion...
but in the food she mixed evil drugs,
that they might utterly forget their native land.

Volume I, Book 10
*Odyssey* by Homer (trans. A. Murray),[69] 8th Century BCE

Odysseus hears of their bewitchment and sets out to rescue them meeting Hermes *en route* who supplies him with an antidote known as "moly", now thought to be *Galanthus nivalis* (snowdrop, Figure 2.5)[70] to protect him from Circe's magic potion.

**Figure 2.5**   Snowdrops in flower, taken in Lincoln.

Although not traditionally used medicinally,[71] the snowdrop contains a chemical called **galanthamine**, an "acetylcholinesterase inhibitor". This means it can restore the action of acetylcholine that scopolamine disrupts, so would be an effective antidote.[72] It is also poisonous itself however, and sometimes the bulbs are mistaken for onions.[10] Like yew (Chapter 3) snowdrops are often found in churchyards,[8] and are considered unlucky to bring into the house.[28] Circe's plant is sometimes depicted in artwork as an apple, which could be the thorn apple (see Chapter 7) rather than the mandrake.[73]

## 2.4   MANDRAKES AND MAGIC

Mandrakes were also associated with magic, with names like "sorcerers' root" and "witches herb". Ben Jonson uses this association with witches (or Hags) in his *Masque of Queens* (1609)[‡]

---

[‡]A masque was a special kind of theatrical performance, usually performed at the Royal Court, in which the audience members themselves would take part in the acting, singing and dancing, often wearing masks to disguise their identity.[22]

3rd HAG:        I lay last night all alone
                  O' the ground to hear the mandrake groan,

*Masque of Queens* by Ben Jonson,[74] 1609

The groaning apparently issued from demonic spirits the roots were said to soak up from the dark earth (although confusingly, the Romans though it could cure demonic possession[75]).[76] Mandrakes were believed to be in league with the Devil[30,77] so their fruits were called "Satan's apple" and "Devil's turnip" as he was said to stand perpetual guard over mandrakes, and if the roots were pulled up at certain holy times, he would appear to do the harvester's bidding, including rejuvenating lost youth.[10,45] It was also said to be very hard to dispose of a mandrake (as it survived drowning, burning and falls from a height) so they tended to be passed down from father to son.[78] In other myths, if you failed to dislodge a mandrake completely when pulling it up (and the roots can extend 1 m into the ground),[71] the earth opened up and you disappeared forever in the grasp of a fiend or fell down straight into Hades.[77,78]

That particular plot device was a little too far-fetched even for the Elizabethans, but another superstition – that mandrakes grow under gallows from the blood and other bodily fluids of the innocent condemned to hang – was used by John Webster's villains in *The White Devil* (1612). Vittoria, a nobleman's daughter, disgraces her family by eloping, and her brothers plot her murder, one of them asking

LODOVICO:      Wilt sell me forty ounces of her blood,
                To water a mandrake?

Act III, scene iii
*The White Devil* by John Webster,[79] 1612

Other poisons play an important role in *The White Devil* as we saw in Chapter 1. In some versions of the superstition, the roots of the mandrake under the gallows extended all the way down to the underworld.[6] If a mandrake was found growing beneath gallows any potions made from it were thought to be extra powerful.[50]

During the 18th and 19th Centuries literary references to the mandrake steadily declined,[30] but two, more modern references to this mandrake myth (where the body fluid alluded to is semen)

can be found in *The Winter of Our Discontent* by John Steinbeck (1961)[80] and Samuel Beckett's play *Waiting for Godot*

VLADIMIR: Where it falls mandrakes grow,
That's why they shriek when you pull them up.
Did you not know that?

Act I
*Waiting for Godot* by Samuel Beckett,[81] 1952

Another legend was that mandrakes grow at crossroads where suicide victims were buried.[47] It is interesting that these myths persisted for so long, given that Turner was trying to dispel them in his herbal as far back as 1568.[24]

Curiously, despite being painted by some as the root of all evil, mandrakes were also thought in medieval times to be effective as an amulet if anyone saw 'great evil in the house'.[71,82] You had to sell your mandrake root before you died for the charm to hold however, and at a lower price than you paid for it.[4]

There are a few superstitions around poppy seeds too. For example, classical gods can be seen in paintings holding or wearing poppies (such as the Roman goddess Ceres and the Greek god Hypnos[8]) and poppy seeds were sometimes placed inside coffins to encourage the body to 'sleep'. They were sometimes used to ward off evil creatures such as vampires as a sort of distraction – apparently on spotting the seeds, the vampire would pause to count them, giving the would-be victim time to escape.[6]

## 2.5 MODERN MEDICAL USES

### 2.5.1 Poppy

Whilst eating[26] and smoking opium has largely fallen out of fashion, it is still harvested as a precursor to **heroin** (prescribed for pain and cough[83]) and used in some Traditional Chinese Medicine (TCM) preparations for things like a chronic (long-term) cough,[84] and to promote digestion.[85] Morphine however is widely used medicinally for pain; in 2020 in England there were more than 5 million prescriptions for morphine on the NHS.[86] It's therefore frequently seen in forensic toxicology cases, as an emergency painkiller in road traffic crashes, as a prescription medication in those receiving palliative care, and alongside other opiates in heroin users who have overdosed (heroin rapidly turns into morphine in the body). For those who accidentally overdose, there is an antidote known as **naloxone**, which reverses the effects of opiates by binding to the receptors instead of the morphine or heroin.

### 2.5.2   Mandrake

Mandrake plants were used in folk medicine to cure constipation (the opposite effect to opiates), and 'exciting the liver to healthy action'.[25] They are still occasionally used in homeopathy[84] or as a poultice or plaster for rheumatic or arthritic pain, or as a decoction (a concentrated liquid made from boiling plant residues) for ulcers.[9] The apples could also be soaked in oil then cooked, filtered and the oil applied to boils.[21] However, it is a restricted herbal ingredient in the UK and *Mandragora autumnalis* can only be made available *via* a prescription from a registered doctor.[87]

Medically, scopolamine absorbs very easily through the skin so is prescribed for motion sickness in the form of a self-adhesive patch that is placed behind the ear (in 2020 in England there were almost 220 000 prescriptions on the NHS for hyoscine for this reason[86]). Scopolamine also deactivates the part of the brain that triggers vomiting, so can be used to treat nausea during chemotherapy and after operations. In the past, scopolamine was given to pregnant women during labour to bring on something called "twilight sleep".[88]

A closely related drug called scopolamine *N*-butylbromide, the ingredient in Buscopan®, which is used in the UK to treat irritable bowel syndrome, turns into scopolamine when it is heated – something it shares with cannabis which also needs to be smoked, baked or vaped to release the main active ingredient $\Delta^9$-tetrahydrocannabinol (THC). Buscopan® is typically prescribed as tablets, but these can be smoked in an attempt to get high, and in Italy in 2021 a 41-year-old male prisoner died suddenly after smoking an unknown number.[89]

### 2.6   MODERN TOXICOLOGY CASES

### 2.6.1   Scopolamine

In 2008 in Norway there was an epidemic of accidental poisonings with scopolamine, which had been added to fake flunitrazepam (Rohypnol®) tablets. The victims presented to hospital with dilated pupils, visual hallucinations, confusion, agitation, and a peculiar plucking behaviour, consisting of seemingly picking up and handling invisible objects from the air and floor.[90] **Physostigmine** (from the *Physostigma venenosum* or Calabar bean plant[91]) was given to some of those suffering scopolamine poisoning, as it acts as an antidote by rapidly reversing the sedating symptoms (another acetylcholinesterase inhibitor) by increasing the amount of acetylcholine in the body.[92]

In 2002, a case of alleged drug-facilitated sexual assault involving scopolamine was reported in Spain.[93] In 2019, a scopolamine-facilitated robbery was reported in France where the assailant blew a powder into the face of the victim.[94] Scopolamine would be an ideal drug for such a purpose; as we have seen, it can lower inhibitions, produce submission through sleep, and cause amnesia.

Poisonings with mandrake plants also still happen; in 2002, a 52-year-old woman was admitted to hospital in Italy after she and five members of her family ate the leaves of a wild plant they believed to be *Borago officinalis*. It was, in fact, mandrake and within hours she became nauseous, confused, and started hallucinating. She was treated by gastric lavage (her stomach was pumped) and recovered within 24 h.[36] It was a lucky escape, as only a few leaves, seeds, berries or pieces of root are needed to produce symptoms of poisoning. In China in 2017 there were two deaths due to scopolamine overdose caused by TCM.[95] These cases can be difficult for toxicologists to investigate as many herbal medicines contain a mixture of chemicals (*e.g.*, *Nao Yang Hua* contains scopolamine, hyoscyamine and atropine).[96] Overdose with herbal TCMs is unfortunately common because decoctions are often prepared at home in rather unscientific ways using varying amounts of water, temperatures, boiling times and plants at different stages of their life cycles.[96]

Normal therapeutic concentrations of scopolamine are tiny: 0.0001–0.0003 mg $L^{-1}$ in serum (the watery part of blood).[97] In two cases where patients were admitted to hospital after scopolamine poisoning, their serum concentrations were 0.00045 and 0.00079 mg $L^{-1}$.[90,98] A 35-year-old man hoping to benefit from mandrake's aphrodisiac effects, ended up in hospital after eating five berries. His blood was negative for alkaloids, but scopolamine was found in his urine.[99] Depending on how quickly samples are taken from a poisoning victim, the chemical culprits can have passed through the blood and been metabolised by the kidneys. This is why analysing more than one sample type in a case is considered good practice. Somewhat confusingly, the sedative and hypnotic drug **methaqualone** has the street name "mandrakes" and was sold combined with the antihistamine **diphenhydramine** under the brand name Mandrax®. Although the symptoms of overdose are similar,[100] methaqualone is a synthetic drug and the two substances are not chemically related.

### 2.6.2 Morphine

Most modern forensic toxicology cases involve people taking pharmaceutical preparations of morphine or scopolamine, but occasionally a toxicologist will see someone who has overdosed accidentally after growing, harvesting and consuming opium poppies. Although poppies can be seen in many gardens in the UK for decoration,[101] growing and processing your own opium poppies is not advisable however, as in the UK, cultivating raw opium is a Schedule 1 offence under the Misuse of Drugs Regulations 2001. *Papaver somniferum* is also another restricted herbal ingredient.

What constitutes a therapeutic concentration of morphine varies hugely between patients by age, route of administration and **tolerance** to its effects, with a range of $0.01–0.12$ mg $L^{-1}$ in serum,[97,98] which has significant overlap with those concentrations associated with toxicity. Tolerance is where more and more of a drug is needed to achieve the same effect[102] – it can make interpretation of forensic toxicology cases difficult as a therapeutic concentration in one person could be toxic or even fatal in another.

In 2009, a 32-year-old man died in Spain after cutting open poppy capsules from a legal poppy field and ingesting an unknown quantity of the latex. His post-mortem blood concentration of morphine was $0.13$ mg $L^{-1}$, but he was also positive for **codeine** and thebaine.[103] As we saw earlier, consuming alcohol and opium together enhances morphine's effects, and in 2014 two deaths in France were reported after users consumed *rachacha* with beer. *Rachacha* is a homemade decoction of poppy heads that is made into a black paste by evaporating off the water, and then swallowed in the form of balls.[104] There are occasional cases of children overdosing after being given a 'calming tea' made from unripe poppy capsules.[5]

## 2.7 CATCHING THE CHEMICAL CULPRIT

In our cases from fiction, there was no proof that the characters had actually ingested morphine or scopolamine. Until qualitative tests were invented in the 19th Century,[105] poisoning diagnoses were made from symptoms, and possibly by examining the victim's stomach contents. These days, scopolamine and morphine can be detected in blood or other body fluids using a technique called **liquid chromatography-mass spectrometry (LC-MS)**. This is really two instruments joined together. In the first instrument, the liquid sample (in our case a cleaned-up body fluid such as blood) is pushed down a long thin tube or "column" under pressure by another liquid (usually a mix of solvents).[106] As the sample travels down the column, any drugs or poisons in it gradually separate out and come out of the end at different times. In the second instrument, the mass spectrometer, we identify the drugs one at a time by looking at the characteristic patterns they make when they are electrically charged (made into positive or negative ions) and broken into tiny pieces. The size of the pieces along with how common each one is compared to others (the "ion ratio") helps us distinguish between drugs that share common ions. In LC-MS we are also interested in how the breakdown happens – which bigger ions fall apart into smaller ones (known as "transitions"). Together, the two techniques can tell us what is present in the sample and how much. Best practice[107] in forensic toxicology is to compare the characteristic patterns to a drug standard or reference material. These are pure, certified drug powders or solutions bought from reputable chemical suppliers. Although we can compare our findings to a 'library' of drugs that have been analysed before, some different drugs can look very alike in these libraries, so running a real sample of the drug alongside the case samples is an extra check.

## 2.8 CASE CLOSED

As we have seen, the ingredients in Barabas' juyce (poppy and mandrake) contain morphine and scopolamine, and both of these drugs can cause drowsiness and sleep. When taken together, they will enhance each other's effects causing even more sedation, and could realistically produce a death-like coma.

To this day, the arms of the Association of Anaesthetists of Great Britain and Ireland contains both poppy heads and the mandrake plant.[17] There was a real case in 1674 in Basingstoke in the UK, where a woman was buried alive after the doctor mistook her poppy-tea-induced sleep for death – the checks were pretty minimal, using a mirror to check if she was breathing before burying her in the local cemetery.[6]

## REFERENCES

1. C. Marlowe, *The Jew of Malta*, Cambridge University Press, Cambridge, 1973.
2. *Gerard's Herbal*, ed. M. Woodward, Senate, London, 1994.
3. A tale of two poppies, https://www.kew.org/read-and-watch/tale-two-poppies, accessed May 2020.
4. S. T. Dietz, *The Complete Language of Flowers: A Definitive and Illustrated History*, Wellfleet Press, New York, 2020.
5. D. Frohne and H. J. Pfänder, *A Colour Atlas of Poisonous Plants*, Wolfe Publishing Ltd, London, 1983.
6. F. Inkwright, *Botanical Curses and Poisons*, Liminal 11, London, 2021.
7. V. Thomas and N. Faircloth, *Shakespeare's Plants and Gardens: A Dictionary*, Bloomsbury, London, 2016.
8. R. Richardson, *Britain's Wild Flowers: A Treasury of Traditions, Superstitions, Remedies and Literature*, The National Trust, London, 2017.
9. A. Chevallier, *Encyclopedia of Medicinal Plants*, DK Publishing, St Leonards, 2001.
10. Anonymous, *The Poison Garden*, The Alnwick Garden, Alnwick, 2005.
11. B. Hubbard, *Poison: The History of Potions, Powders and Murderous Practitioners*, Welbeck, London, 2020.
12. M. R. Cooper, A. W. Johnson and E. A. Dauncey, *Poisonous Plants and Fungi: An Illustrated Guide*, The Stationery Office, Norwich, 2nd edn, 2003.
13. M. Castro, *The Complete Homeopathy Handbook: A Guide to Everyday Health Care*, Macmillan, London, 1990.
14. M. Walker, *Why We Sleep*, Penguin, London, 2018.
15. P. M. North, *Poisonous Plants and Fungi*, Blandford Press, London, 1967.

16. K. Dettmer, B. Saunders and J. Strang, *Br. Med. J.*, 2001, **322**, 895–896.
17. A. J. Carter, *Br. Med. J.*, 1996, **313**, 1630–1632.
18. The Royal Botanical Gardens Kew, *Plants+People*, Kew Publishing, London, 1998.
19. J. Gerard and T. Johnson, *The Herball or Generall Historie of Plantes*, Adam Islip, Joice Norton & Richard Whitakers, London, 2nd edn, 1636.
20. Friedrich Sertürner, https://webarchive.nationalarchives. gov.uk/20150208094122/http://www.sciencemuseum. org.uk/online_science/explore_our_collections/people/ serturner_friedrich, accessed May 2020.
21. M. Brown, *Death in the Garden: Poisonous Plants & their Use Throughout History*, Pen & Sword Books Ltd, Barnsley, 2018.
22. K. Harkup, *Death by Shakespeare: Snakebites, Stabbings and Broken Hearts*, Bloomsbury, London, 2020.
23. M. Grieve, *A Modern Herbal*, Dover Publications, New York, 2nd edn, 1971.
24. W. Turner, *A New Herball*, Cambridge University Press, Cambridge, 1989.
25. M. Chamberlin, *Old Wives' Tales: The History of Remedies, Charms and Spells*, The History Press, Cheltenham, 2020.
26. L. S. Nelson, M. A. Howland, N. A. Lewin, S. W. Smith, L. R. Goldfrank and R. S. Hoffman, *Goldfrank's Toxicologic Emergencies*, McGraw Hill, New York, 11th edn, 2018.
27. M. Blamey and C. Grey-Wilson, *Mediterranean Wild Flowers*, Harper Collins, London, 1993.
28. F. Inkwright, *Folk Magic and Healing: An Unusual History of Everyday Plants*, Liminal 11, London, 2019.
29. L. O. Hanuš, T. Řezanka, J. Spížek and V. M. Dembitsky, *Phytochemistry*, 2005, **66**, 2408–2417.
30. M. R. Lee, *J. R. Coll. Physicians Edinburgh*, 2006, **36**, 278–285.
31. T. E. Keys, *The History of Surgical Anesthesia*, Dover Publications, New York, 1963.
32. G. Chaucer, *Troilus and Cressida and The Canterbury Tales*, Willam Benton, Chicago, 1952.
33. A. Bassolii, G. Borgonovo and G. Busnelli, in *Modern Alkaloids*, ed. E. Fattorusso and O. Taglialatela-Scafati, Wiley, New York, 2007, ch. 3, pp. 53–72.
34. D. F. Duarte, *Rev. Bras. Anesesiol.*, 2005, **55**, 135–146.

35. W. Shakespeare, *Cymbeline*, Pelican, New York, 1979.
36. G. A. Piccillo, E. G. M. Mondati and P. A. Moro, *Eur. J. Emerg. Med.*, 2002, **9**, 342–347.
37. A. J. Carter, *J. R. Soc. Med.*, 2003, **96**, 144–147.
38. J. Robertson, *Is That Cat Dead?*, Book Guild Publishing, Brighton, 2010.
39. W. Shakespeare, in *Comedies*, David Campbell Publishers Ltd, London, 2000, ch. 5, vol. 1, pp. 408–511.
40. J. P. André, *J. Chem. Educ.*, 2013, **90**, 352–357.
41. A. S. Harper-Leatherman and J. R. Miecznikowski, *J. Chem. Educ.*, 2012, **89**, 629–635.
42. J. K. Rowling, *Harry Potter and the Goblet of Fire*, Bloomsbury, London, 2000.
43. J. K. Rowling, *Harry Potter and the Chamber of Secrets*, Bloomsbury, London, 1998.
44. W. Shakespeare, *Antony & Cleopatra*, Cambridge University Press, Cambridge, 1950.
45. T. F. Thiselton Dyer, *Folk-Lore of Shakespeare*, Dover Publications, New York, 1966.
46. W. Shakespeare, *Othello*, Wordsworth Editions Ltd, Ware, 1997.
47. K. Fatur, *Econ. Bot.*, 2020, **20**, 1–19.
48. Apuleius, *The Golden Ass or Metamorphoses*, Penguin, London, 2004.
49. E. A. Campbell, *Victorian Lit. Cult.*, 2007, **35**, 607–615.
50. A. C. Kail, *Med. J. Aust.*, 1983, **2**, 515–519.
51. M. Kobs, MA thesis, University of Missouri, 2009.
52. J. Timbrell, *The Poison Paradox: Chemicals as Friends and Foes*, Oxford University Press, Oxford, 2005.
53. J. Mann, *Murder Magic and Medicine*, Oxford University Press, Oxford, 1992.
54. J. K. Rowling, *Harry Potter and the Deathly Hallows*, Bloomsbury, London, 2007.
55. W. Shakespeare, *The Second Part of King Henry VI*, Cambridge University Press, Cambridge, 1991.
56. T. Moore, *Lalla Rookh*, George Routledge & Sons, London, 1877.
57. P. Dioscorides, *The Greek Herbal of Dioscorides*, Haner Publishing, New York, 1959.

58. *The New Oxford Annotated Bible (New Revised Standard Version)*, ed. M. D. Coogan, M. Z. Brettler, C. A. Newsom and P. Perkins, Oxford University Press, Oxford, 2006.
59. J. M. Riddle, *Goddesses, Elixirs, and Witches*, Palgrave MacMillan, New York, 2010.
60. R. Young, *Young's Analytical Concordance to the Bible*, W. M. B. Eerdmans Publishing Co., Grand Rapids, MI, 1975.
61. W. Shakespeare, *The History of King Henry the Fourth Part 2*, Clarendon, Oxford, 1998.
62. T. A. Sumberg, *J. Polit.*, 1961, **23**, 320–340.
63. J. Crawford, MA thesis, The University of Alabama, 2017.
64. M. Kaplan, in *Discover Magazine*, Kalmbach Media, Waukesha, 2015.
65. E. Tabor, *Econ. Bot.*, 1970, **24**, 81–94.
66. W. Shakespeare, *Macbeth*, Cambridge University Press, Cambridge, 1960.
67. J. Webster, *The Duchess of Malfi*, Oliver and Boyd, Edinburgh, 1972.
68. F. Khakpai, M. Nasehi, A. Haeri-Rohani, A. Eidi and M. R. Zarrindast, *Behav. Brain Res.*, 2012, **231**, 1–10.
69. Homer, *Odyssey*, Harvard University Press, Cambridge, MA, 1995.
70. A. Plaitakis and R. C. Duvoisin, *Clin. Neuropharmacol.*, 1983, **6**, 1–5.
71. M. Grieve, *A Modern Herbal*, Tiger Books International, Twickenham, 3rd edn, 1998.
72. J. Emsley, *Vanity, Vitality, and Virility: The Science Behind the Products You Love to Buy*, Oxford University Press, Oxford, 2004.
73. R. Ashton and C. Leblanc, *Dalhousie Med. J.*, 2010, **37**, 29–31.
74. B. Jonson, *The Masque of Queens*, Yale University Press, New Haven, CT, 1969.
75. A. Stewart, *Wicked Plants*, Algonquin Books of Chapel Hill, Chapel Hill, NC, 2009.
76. T. Hargreaves, *Poisons and Poisonings: Death by Stealth*, RSC Publishing, Cambridge, 2017.
77. S. Lawrence, *Witch's Garden: Plants in Folklore, Magic and Traditional Medicine*, Welbeck, London, 2020.

78. C. M. Skinner, *Myths and Legends of Flowers, Trees, Fruits, and Plants: In All Ages and in All Climes*, J. B. Lippincott, Philadephia, 1911.

79. J. Webster, *The White Devil*, Methuen & Co. Ltd, London, 1965.

80. J. Steinbeck, *The Winter of Our Discontent*, Penguin, London, 2000.

81. S. Beckett, *Waiting for Godot: A Tragicomedy in two Acts*, Faber and Faber, London, 2006.

82. A. Van Arsdall, *Medieval Herbal Remedies: The Old English Herbarium and Anglo-Saxon Medicine*, Routledge, New York, 2002.

83. A. Foster, *The Medicinal Plant Collection at the University of Oxford Botanic Garden*, Wellcome Trust, Oxford, 2010.

84. Medical Economics Company, *PDR for Herbal Medicines*, Medical Economics Company, Montvale, NJ, 1998.

85. J. Zhou, G. Xie and X. Yan, *Encyclopedia of Traditional Chinese Medicines*, Springer, Heidelberg, 2011.

86. Prescription Cost Analysis - England, 2020, https://www.nhsbsa.nhs.uk/statistical-collections/prescription-cost-analysis/prescription-cost-analysis-england-2019, accessed June 2021.

87. Banned and restricted herbal ingredients, https://www.gov.uk/government/publications/list-of-banned-or-restricted-herbal-ingredients-for-medicinal-use/banned-and-restricted-herbal-ingredients, accessed May 2020.

88. J. H. Bock and D. O. Norris, in *Forensic Plant Science*, ed. J. H. Bock and D. O. Norris, Academic Press, San Diego, 2016, ch. 1, pp. 1–22.

89. S. Strano-Rossi, S. Mestria, G. Bolino, M. Polacco, S. Grassi and A. Oliva, *Int. J. Leg. Med.*, 2021, **135**, 1455–1460.

90. O. M. Vallersnes, C. Lund, A. K. Duns, H. Netland and I.-A. Rasmussen, *Clin. Toxicol.*, 2009, **47**, 889–893.

91. W. Sneader, *Drug Discovery: A History*, John Wiley & Sons, Chichester, 2005.

92. A. Chadwick, A. Ash, J. Day and M. Borthwick, *BMJ Case Rep.*, 2015, **2015**, bcr2015209333.

93. A. de Castro, E. Lendoiro, Ó. Quintela, M. Concheiro, M. López-Rivadulla and A. Cruz, *Forensic Toxicol.*, 2012, **30**, 193–198.

94. L. Dufayet, E. Alcaraz, J. Dorol, C. Rey-Salmon and J.-C. Alvarez, *Forensic Toxicol.*, 2020, **38**, 264–268.
95. M. Pan, X. Wang, Y. Zhao, W. Liu and P. Xiang, *Forensic Sci. Int.*, 2019, **298**, 39–47.
96. N. K. Ho, *Singapore Med. J.*, 2001, **42**, 487–492.
97. *Clarke's Analysis of Drugs and Poisons*, ed. A. C. Moffatt, D. Osselton and B. Widdop, Pharmaceutical Press, London, 4th edn, 2011.
98. M. Schulz, A. Schmoldt, H. Andresen-Streichert and S. Iwersen-Bergmann, *Crit. Care*, 2020, **24**, 195.
99. P. Nikolaou, I. Papoutsis, M. Stefanidou, A. Dona, C. Maravelias, C. Spiliopoulou and S. Athanaselis, *J. Emerg. Med.*, 2012, **42**, 662–665.
100. D. Baggish, S. Gray, P. Jatlow and M. J. Bia, *Yale J. Biol. Med.*, 1981, **54**, 147–150.
101. R. Bevan-Jones, *Poisonous Plants: A Cultural and Social History*, Windgather Press, Oxford, 2009.
102. J. Keogh, *Pharmacology*, McGraw-Hill, New York, 2010.
103. M. A. Martínez, S. Ballesteros, E. Almarza and J. Garijo, *Forensic Sci. Int.*, 2016, **265**, 34–40.
104. C. Monteil-Ganiere, J.-M. Gaulier, D. Chopineaux, L. Barrios, A. Pineau, É. Dailly and R. Clément, *Forensic Sci. Int.*, 2014, **245**, e1–e5.
105. A. Pappas, N. Massoll and D. Cannon, *Ann. Clin. Lab. Sci.*, 1999, **29**, 253–262.
106. J. Emsley, *Molecules of Murder: Criminal Molecules and Classic Cases*, RSC Publishing, Cambridge, 2008.
107. S. P. Elliott, D. W. S. Stephen and S. Paterson, *Sci. Justice*, 2018, **58**, 335–345.

# A Ghostly Encounter with Poison

If you see a term that's **bold** it's defined in the Glossary. Only the first time that the word appears in the chapter will it be indicated in this way.

---

**Case History: Sudden Death**

*Hamlet Jnr is visited by a ghost claiming to be his father, Hamlet Snr, former King of Denmark, who reveals that his brother Claudius murdered him by pouring a poison in his ear.*

GHOST:    ...Sleeping within my orchard,
             My custom always in the afternoon,
             Upon my secure hour thy uncle stole
             With juice of cursèd hebenon in a vial,
             And in the porches of my ears did pour
             The leperous distilment, whose effect
             Holds such an enmity wi'th' blood of man...

                                    Act I, scene v

*The Tragedy of Hamlet, Prince of Denmark* by William Shakespeare,[1] 1600

---

Poisonous Tales: A Forensic Examination of Poisons in Fiction
By Hilary Hamnett
© Hilary Hamnett 2023
Published by the Royal Society of Chemistry, www.rsc.org

## 3.1 THE INVESTIGATION

Is it plausible that the king's death could be caused in this way? At least 1500 years before Shakespeare's time, Pliny's *Natural History* contained a recipe for curing earache containing henbane, opium and rose oil, which was administered by a syringe into the ear.[2,3] And by the 16th Century this had made it into the medical textbooks of the time, including in an anonymous work known (after the publisher) as *Banckes's Herbal*.[4,5] Since then, modern medicine has confirmed that a number of poisons and medicines can be, and are, absorbed through the thin membranes inside the intact ear.[6] If you have ever had an inner ear infection, you may have been given ear drops for example, and folk remedies for earache involved placing drops of almond oil in the ear.[7]

In our *Hamlet* case history, the "juice" is described as a "distilment", which is most likely an alcoholic distillation (purified liquid) of essential plant oils.[6] The plant is named in the play as "hebenon" or "hebona" depending on which version of *Hamlet* you are reading. It's a fictional plant, the identity of which has been subject to much speculation, but three of the candidates we will consider in this chapter are henbane,[4,8,9] hemlock and yew.[10] Another candidate – ebony – has also been suggested,[11] but ebony is not poisonous nor even psychoactive.[†] A fourth possibility is deadly nightshade,[12] which we will meet in the next chapter.

Hamlet Snr's symptoms (apart from death) are not described in detail, but Shakespeare writes

GHOST:    That swift as quicksilver it courses through
          The natural gates and alleys of the body,
          And with a sudden vigour it doth posset
          And curd, like eager droppings into milk,
          The thin and wholesome blood. So did it mine;
          And a most instant tetter barked about,
          Most lazar-like, with vile and loathsome crust,
          All my smooth body.

Act I, scene v
*The Tragedy of Hamlet, Prince of Denmark*
by William Shakespeare,[1] 1600

---

[†]Ebony was easily confused with guaiac wood from the *Guaiacum officinale L.* tree – a source of a treatment for syphilis or pox.

We know the speed of onset of the poison's symptoms was rapid or "with a sudden vigour". The "curd" could indicate blood coagulation, and the "most instant tetter barked about" may have been a skin disease such as dermatitis.[6] The phrase "lazar-like" could be a reference to the biblical character of Lazarus (Luke 16:20) who suffered from leprosy and was covered in sores.[13] Whatever the poison may have been, Shakespeare's description has been interpreted as a rapid paralysis of respiration and circulation.[6]

## 3.2  THE PLANTS BEHIND THE STORY

So do these symptoms fit with any of our poisonous candidates?

### 3.2.1  Henbane

Henbane has a close match in spelling with hebenon, and in fact Christopher Marlowe also used a similar word—hebon—in his play *The Jew of Malta* (1589–90), which appeared 10 years before *Hamlet*. In this play, which we met in Chapter 2, the scheming Barabas describes a poison he is planning to use on a convent of nuns that includes his own daughter

BARABAS:       ...As fatall be it to her as the draught
               Of which great *Alexander* drunke, and dyed:
               And with her let it worke like *Borgias* wine,
               Whereof his sire, the Pope, was poysoned.
               In few, the blood of Hydra, Lerna's bane;
               The jouyce of Hebon, and *Cocitus* breath,

                                                      Act III, scene iv
                            *The Jew of Malta* by Christopher Marlowe,[14] 1589–90

Henbane and hemlock also feature in the *Masque of Queens* (1609) we met in Chapter 2, and are used by the witches to make an exceptionally poisonous brew. Another of the ingredients – nightshade – features in Chapter 4

9th HAG:       And I ha' been plucking, plants among,
               Hemlock, henbane, adder's tongue,
               Nightshade, moonwort, libbard's bane,

                            *Masque of Queens* by Ben Jonson,[15] 1609

In the Ancient Greek epic poem *Argonautica* by Apollonius Rhodius (3rd Century BCE), which tells the tale of Jason and the Argonauts, Medea the witch gives him a potion to protect him from pain during his efforts to seize the golden fleece

She [Medea], meanwhile, took from the hollow chest a drug
which they say is called Promethean
If ... a man should anoint his body with this drug,
he would truly be impervious to strokes of bronze
and not yield to blazing fire,
but for that day would be superior both in valor and might.

Lines 844–850
Book III
*Argonautica* by Apollonius Rhodius (trans. W. H. Race),[16]
3rd Century BCE

The "drug" here is a herb that Jason soaks in water and rubs on his body and weapons,[17] and is thought by some to be henbane,[18] while others have suggested it was mandrake (Chapter 2).[19]

Henbane (*Hyoscyamus niger*) also known as "black henbane", "sleeping herb", "fetid nightshade", or "Devil's eye" has yellow or green funnel- or trumpet-shaped flowers with purple veins (Figure 3.1).[20-26] Henbane springs out of the ground in May and the flowers appear in August,[24,27] and is hugely popular with snails, who seem able to withstand its poisonous effects.[27] The stems of mature plants are covered with long glandular hairs, are sticky,[28] and produce thousands of tiny black seeds (similar to poppy seeds) in October.[24,29,30] It is also called "stinking night-shade" because of its foul fishy smell, which is caused by a compound related to **putrescine**, the odour of death.[29,31] Some of its nicotine-like effects have led it to be called "Poison tobacco".[32]

Henbane contains more than 30 chemicals or alkaloids, the most important of which are **hyoscine** (also known as **scopolamine**) and **hyoscyamine**.[10] Different chemicals are found in different parts of the plant – scopolamine is more commonly found in the green tops and leaves,[20] and hyoscyamine in the seeds.[30] To make using the plant even more risky, there is inter-conversion between hyoscyamine, **atropine** and scopolamine during storage.[32] Compared to deadly nightshade, which we will meet in Chapter 4, and Datura (Chapter 7) the amounts of alkaloids are

**Figure 3.1** A henbane plant in flower, taken at the Oxford Botanic Garden and Arboretum.

very variable in henbane plants, making any remedies made from it highly unpredictable.[33]

Known since Ancient Greece, henbane was used as a salve for pain and to help with insomnia, bringing on prolonged unconsciousness.[34] It was also used as an anaesthetic by 'doctors' at travelling fairs who used the fumes of burning henbane seeds (or the roots mixed with vinegar) to cure toothache,[3,29] possibly because of the resemblance of the plant to a jaw with molars.[20,24] Of course, the 'cure' was short-lived and the toothache returned, usually once the fair had left town. A possibly related wheeze was selling a kind of necklace made from henbane roots to dangle on the necks of children for easy teething.[30] It's not clear whether these were magical amulets or whether they had a placebo or genuine topical pain-relieving effect, because as we saw in Chapter 2, **scopolamine** absorbs easily through the skin.[27]

Henbane seeds could also be pressed into oil and rubbed on the skin, apparently to cause hair to grow more slowly.[19] In classical times it was made into suppositories to relieve pain,[35] and the seeds were smoked in pipes to cure rheumatism.[36]

Tinctures or plant extracts of henbane dissolved in alcohol have also been used to treat urinary tract problems,[37] and even as a contraceptive.[38] It has been known since Ancient Egyptian times that plant extracts dissolved better in alcohol than water.[39] Another use for henbane (and the mandrakes we met in the previous chapter) was to fortify beer by enhancing the effects of the alcohol, although this was eventually outlawed.[29]

Use of moderate doses of henbane can bring on nausea, hallucinations, alterations in heartbeat[30] and fatigue.[32] But in high doses, henbane can be fatal, and was used for executions and poisonings in upper class 17th Century France.[8] Lethal doses initially bring on CNS stimulation followed by asphyxiation[32] and, as Gerard tells us, a deadly sleep.[24,34] Another botanical author of Shakespeare's time, Turner, wrote that henbane "makes men mad, and fall into a great sleep".[3]

As we will see with many poisonous plants, the line between therapeutic use and toxicity is a fine one. Because the potency of plants depends on how and where they are grown, and there was no accurate way of measuring dose, many accidental poisonings happened. Henbane's alkaloids also withstand boiling and drying,[37] and the combination of chemicals works in synergy, enhancing each other's effects.[40] It wasn't until the 1880s when scientists isolated the chemicals responsible for henbane's effects that it could be used safely.

The symptoms recorded for henbane poisoning in the herbals are similar but not identical to those of hebenon in *Hamlet*.[4] One similarity is that the onset of symptoms including pain is rapid,[41] and henbane could have been absorbed through the ear. It has been known for hundreds of years that mixing plant constituents such as henbane with fats or oils makes them easier to absorb through the skin,[10] and in fact henbane extract was widely used in the ears as a treatment for deafness and earache.[6,10]

### 3.2.2   Hemlock

Hemlock was also known during Shakespeare's time,[41] and the three witches (Figure 3.2) in *Macbeth* (1605) use it as an ingredient in their potion

**Figure 3.2**   The Three Witches from Shakespeare's Macbeth, by Daniel
Gardner.

THIRD WITCH:          Scale of dragon, tooth of wolf,
                     Witches' mummy, maw and gulf
                     Of the ravin'd salt-sea shark,
                     Root of hemlock digg'd i' the dark,

Act IV, scene i
*The Tragedy of Macbeth* by William Shakespeare,[42] 1605

Collecting plants in the dark was quite common in Shakespeare's day – partly to avoid being seen with something poisonous, but also because plants were thought to be at their most
potent at night.[43] Gerard warns us even back in the 1630s that
hemlock should not be used for medicinal purposes and that
it "hurteth the heart and liver". Although he and Turner also
suggest drinking hot wine as an antidote to hemlock before
the venom has taken hold of the heart.[3,44] We now use the term
"venom" to describe something injected by an animal or insect
into the body (see Chapter 9) and refer to hemlock as containing
a "poison".

In *King Lear* (1605), the king's rejected youngest daughter Cordelia laments his madness to a doctor, by describing him as missing and wondering around wearing a crown made of hemlock. The crown representing the inversion of majesty.[45] As we will see, hemlock can cause the appearance of madness (eye rolling and drooling) but also some psychological symptoms such as hallucinations

CORDELIA:     Alack, 'tis he: why, he was met even now,
              As mad as the vexed sea, singing aloud,
              Crowned with rank fumitor and furrow-weeds,
              With burdocks, hemlock, nettles, cuckoo-flowers,

Act IV, scene iii
*The Tragedy of King Lear* by William Shakespeare,[46] 1605

A poison drink containing various plants features in the opera *Suor Angelica* (translated as *Sister Angelica*) by Giacomo Puccini, which premiered in 1918. Sister Angelica who has been isolated in a convent since the birth of her son, learns that he has died of a fever and decides to take her own life by drinking a poison distilled from flowers picked from the convent garden (we explored in Chapter 1 why religious orders had gardens growing poisonous plants).[47] She tells us the story during the aria called Amici fiori

| SUOR ANGELICA: | SISTER ANGELICA: |
| --- | --- |
| Suor Angelica ha sempre una ticetta buona fatta coi fiori. | Sister Angelica was always good at making things from flowers. |
| Amici fiori che piccol seno racchiudete le stille del veleno. | These friendly flowers hide drops of poison in their hearts. |
| Ah, quanto cure v'ho prodigate. | I took so much care with them, and now they will repay me. |
| Per voi, miei fior, io morirò! | Through you, flowers I shall die! |

Amici fiori
*Suor Angelica* by Giacomo Puccini (ed. Burton Fisher),[48] 1918

The flowers have been named as oleander (which we will meet in Chapter 7), cherry laurel, hemlock and belladonna (see Chapter 4).[9] This would certainly be a poisonous cocktail, and we will see later in this chapter that hemlock is still used in suicide attempts.

Keats immortalised hemlock and its effects in his rather gloomy poem about death, *Ode to a Nightingale* (1820)

My heart aches, and a drowsy numbness pains
My sense, as though of hemlock I had drunk,
Or emptied some dull opiate to the drains

*Ode on Melancholy* by John Keats,[49] 1820

Hemlock (*Conium maculatum*) is also known as "spotted hemlock", "bad man's oatmeal", "Devil's porridge" or "poison parsley" because of its resemblance to the non-poisonous variety. It has clusters of flowers with very small white petals that all grow out from a central point (Figure 3.3, left).[32,50] The berries are tiny and oval in shape,[32] and can be mistaken for aniseed or caraway,[51] whilst the seeds have beaded ridges.[52] The leaves (Figure 3.3, right) can be mistaken for lettuce.[53] In myth and legend the purple-streaked stems apparently represent the blood from the

**Figure 3.3**   (Left) A hemlock plant in flower, taken in Lincoln. (Right) Leaves of the hemlock plant, taken at the Chelsea Physic Garden.

biblical murder of Abel by his brother Cain polluting the ground (Genesis 4: 8).[54] The roots also look very similar to parsnip, celery or fennel.[51,55–58] When wilting, the highly poisonous herb apparently smells of mice[32] or urine,[51,59] or according to some sources, mouse urine.[21,60] The plant also tastes disgustingly salty and is pungent particularly when crushed.[32] Although hemlock is one of the first plants we see in Spring in the UK[51] growing alongside fences and roads,[59] when used medicinally it is gathered from June to September.[24,32,61]

There are traditions that hemlock should be gathered at night ("digg'd i' the dark") as this increases its potency.[61] There's no scientific reason why this might be case, so it's likely just a cautionary myth that developed to protect those picking it from being accused of witchcraft.

Poison hemlock is sometimes mistaken for water hemlock (*Cicuta maculate*, Figure 3.4), which is found in swampy areas and although a different species is also poisonous. Water hemlock contains a different toxic chemical, enanthotoxin, which is

**Figure 3.4**    A water hemlock plant in flower, taken at the Chelsea Physic Garden.

an isomer of **cicutoxin**.[51,62] The effects are also different, with water hemlock producing a sudden onset of vomiting.[63] Other plants can look similar to hemlock, such as cow parsley, and in folklore this was called "mother-die" or "stepmother's blessing" and children were warned that if they picked the flowers, or used the stems as whistles[36] something bad would happen to their parents. It's possible that the legend was invented by parents trying to keep their children away from poison hemlock.[64]

Poison hemlock contains 13 chemicals known as **piperidine alkaloids**,[65,66] which look chemically similar to **nicotine** and the most deadly of which is **coniine**,[51,67] but also includes ***N*-methylconiine** and **gamma-coniceine**.[32,68] Although all parts of the plant contain coniine,[51] it is most likely to be present in harmful quantities in the freshly harvested plant, particularly in its roots, berries (which contain an inner wall called the "coniine layer"[59]) and seeds.[32,51] gamma-Coniceine is found in the flower buds and flowers, but is transformed during the fruit development into coniine and then into *N*-methylconiine in the fruit.[65] As coniine is very volatile (evaporates easily) dried plants contain less than fresh plants.[53]

In folk medicine, hemlock was used internally to treat herpes,[51,65] neuralgia, rheumatism of the muscles and joints, stiffness of the neck, cramps, bronchial spasms and whooping cough.[51] It was thought to make breathing easier by relaxing the muscles of the windpipe.[51] It was even once considered an antidote to **strychnine** (see Chapter 8)[66] and an appropriate treatment for teething in children![51] When applied externally it was used as an ointment for coughs, asthma, sciatica, backache[32] or as a plaster.[51] The plasters were applied by monks to the pubic area apparently to quench lustful thoughts.[43] In classical times it was used as a salve for herpes, to dry up breastmilk after birth, and was said to weaken the genitals if smeared on them, reducing wet dreams.[35] It was also possible to make tea out of hemlock leaves[51] as a cure for scurvy. In medieval times it was thought that hemlock could cure the bite of a mad (rabid) dog,[51] by mixing hemlock seeds with fennel in wine.[65]

Use of moderate doses of hemlock can bring on headache, burning of the mouth, dilated pupils, salivation, rolling of the eyes, visual disorders, jelly legs, and mydriasis.[20,32,51,69] It is also a violent emetic, causing both vomiting and diarrhoea.[70] Symptoms typically commence in 15 min.[63]

In higher doses, hemlock can be fatal because coniine and gamma-coniceine act to block the transmission of nerve impulses.[51,71] Initially this briefly stimulates the CNS, causing dizziness, hallucinations and convulsions.[29,51] After this, the alkaloids close down the impulses related to feeling and movement, resulting in paralysis of the muscles. Paralysis starts from the feet and moves upwards,[32,51,71] eventually spreading to the chest[51] and bringing the heart to a halt.

Again we have seen in real life some of the same symptoms as Hamlet senior describes, including cases of dermatitis[69] and we know that coniine absorbs through the skin[51] and mucous membranes.[32]

A recipe for a kind of ball of narcotic plant material called a "sleeping apple" from classical times included opium and mandrake (Chapter 2), hemlock and henbane. Apparently even just holding the 'apple' could make someone fall unconscious, and it could be that this provided the inspiration for the poisoned apple in *Snow White*.[27] Another possibility is the cyanide-containing compounds found in apple pips, which we will look at in more detail in Chapter 10.

### 3.2.3  Yew

The yew tree has a long history of association with death and mourning, and was carried by the Ancient Greeks in funeral processions.[54] It was known to be poisonous in Shakespeare's time, with Gerard telling us that the yew is "very venomous to be taken inwardly".[44] Shakespeare uses many of the yew myths in his plots with the first being its poisonous use in magic – it was another ingredient in the witches' brew we saw in *Macbeth* earlier, with the "slips" likely to be leaves or needles from the yew tree (Figure 3.5)

THIRD WITCH:      Gall of goat, and slips of yew
                  Silver'd in the moon's eclipse,

Act IV, scene i
*The Tragedy of Macbeth* by William Shakespeare,[42] 1605

In *King Richard II* (1595), the king banishes his rival Henry Bullingbrook, seizes his land, and uses the money to fund a war in Ireland. On his return, he seeks refuge in a Welsh castle and

**Figure 3.5**   (Left) An Irish yew, taken at the Chelsea Physic Garden. (Middle) An English yew, taken at the Cambridge University Botanic Garden, and (right) close-up of yew needles and berries, taken in Lincoln.

Scroope, one of his few remaining supporters, describes how the tide has turned against them

SIR STEPHEN
SCROOPE:            ...In stiff unwieldy arms against thy crown.
                    The very beadsmen learn to bend their bows
                    Of double-fatal yew against thy state.

                                                        Act III, scene ii
                        King *Richard II* by William Shakespeare,[72] 1595

The "double-fatal" referring to the fact that yew is poisonous and the wood was often used to make the long-bows of archers because of its strength and resistance to water.[36,61,73]

In *Romeo & Juliet* (1594), Paris (Romeo's love rival) and a servant boy are in a churchyard near to the tomb of Juliet's family, the Capulets, where Juliet is feigning death thanks to a potion (see Chapter 4). Paris leaves his servant hiding under a yew tree to act as a lookout before entering the tomb himself

PARIS:     Give me thy torch, boy: hence, and stand aloof:
           Yet put it out, for I would not be seen.
           Under yond yew-trees lay thee all along,
           Holding thine ear close to the hollow ground;
           So shall no foot upon the churchyard tread,

                                                        Act V, scene iii
            *The Tragedy of Romeo & Juliet* by William Shakespeare,[74] 1594

Yew trees are still a common sight in English churchyards[24] and cemeteries (even going by the name "graveyard tree"[75] although this custom predates Christianity[76]), and we meet one in such a location near the end of *Harry Potter and Goblet of Fire* (2000).[77] The roots of the yew are very fine and so a legend started that they would grow through the eyes of the dead buried below to prevent them seeing their way back to the World of the living.[70] The roots were also believed to drink up the poisonous exhalations of the dead,[54] or grow through their mouths, freeing the souls for rebirth.[76]

This role of the yew features in the Tennyson poem *In Memoriam* (1850)

Old Yew, which graspest at the stones
That name the under-lying dead,
They fibres net the dreamless head,
Thy roots are wrapt about the bones.

2
*In Memoriam A.H.H.* by Alfred Tennyson,[78] 1850

Lying under a yew tree was a risky business however, as another common legend was that sleeping under its shadow could cause death or sickness.[24] Although just being near a yew tree is not dangerous, this myth was probably designed to keep people away from its tempting berries. The unluckiness of the yew features in another of Shakespeare's plays, *Titus Andronicus* (1593). Roman general Titus returns from war with Tamora, queen of the Goths and her three sons as prisoners, all of whom vow to take revenge against him. As part of a plot to have her sons kill the Emperor's brother and then frame someone else for the murder, Tamora persuades them her life is in danger

TAMORA:      ...No sooner had they told this hellish tale
             But straight they told me they would bind me here
             Unto the body of a dismal yew
             And leave me to this miserable death.

Act II, scene iii
*Titus Andronicus* by William Shakespeare,[79] 1593

The phrase "stuck all with yew" sung by a jester in *Twelfth Night* (1599),[80] means that after death, the body was adorned with sprigs of the yew tree. This was common practice in mourning rituals, where yew wreaths or crowns were placed on bodies and even on the heads of animals sacrificed to some gods.[73]

Yew (*Taxus baccata* fastigiata) also known as the "tree of death" or "Irish yew", is an upright evergreen shrub (Figure 3.5, left) with rust-coloured bark, dark green glossy leaves, small yellowish flowers and red berries called "arils" containing a single seed (Figure 3.5, right).[20,21,52,81] It is often manicured into hedges or ornamental features, unlike the common or "English" yew (from which it descends), which grows into large spreading trees up to 20 m tall (Figure 3.5, middle) and can survive thousands of years.[28] Both are poisonous, but unfortunately authors of fiction rarely tell us which one they mean. Irish yew trees are apparently all descended from cuttings of one tree in Ireland, making them all female relations.[27]

Unlike many of the plants we have seen, it is the needles rather than the berries (Figure 3.5, right) that are the most poisonous part of the yew tree (along with the seeds and bark).[28,37] Unfortunately, they look very similar to rosemary leaves.[43] They are deadly because yew contains a group of chemicals known as "taxine alkaloids" or "taxines", the two major ones being imaginatively named **taxine A** and **taxine B**, but others include **taxicatine**.

Small doses of yew leaves have been used to promote menstruation, to treat diptheria, epilepsy, tapeworm and tonsillitis.[32] When more are ingested (and the lethal dose for an adult is reported to be 50–100 g of yew needles[32,82]) symptoms include impaired colour vision, vomiting, diarrhoea, abdominal pain, delirium, convulsions, unconsciousness, and a dangerous drop in pulse rate, which ultimately stops the heart, leading to cardiac arrest. Survival after poisoning is rare.[32,37,82] **Taxine B** is responsible for the greatest cardiac toxicity by increasing calcium levels in cells.[82]

So could hebona be yew? Its poisonous nature was known in Shakespeare's time, even if it went by the name "Ughe" in some herbals.[3] Although the symptoms are indeed similar to some of Hamlet Snr's, there is not much historical evidence of yew extracts being absorbed through the skin or the ear.[45] Most of the recorded cases of poisoning come from eating yew needles and the speed at which symptoms are noticed is slow (30 min to 1 h). The time to death is even longer, taking up to 5 h.[82]

## 3.3 HAGS, HENBANE AND HEMLOCK

### 3.3.1 Henbane

As we have seen from the literary references to our three poisonous candidates, they were all associated with witchcraft. Henbane was known as a favourite of witches – hags were thought to burn the plant, and inhale the fumes conjuring up the spirits and demons they needed to perform magic spells and brew up potions.[29] Henbane and mandrake (which we met in the previous chapter) were used together in witches' "flying ointments" or "salves" to produce hallucinations,[83–85] and also in love potions.[70] The dead in Hades were said to wear crowns of henbane's "unsanctified looking" flowers as they wandered hopelessly beside the River Styx.[54] In Victorian times, henbane was said to mean "defect", "absence" or "imperfection".[86] In *Hamlet*, the play-within-a-play performed by a band of travelling actors, which re-tells the story of the King's death, has the poison being prepared by witches.[87]

Along with witches, another group said to have used psychoactive substances were the fierce Norse berserker warriors. They were said to enter a special trance before battle, which may have been due to a few different plant culprits. Traditionally the plants were thought to be the mushrooms *Amanita muscaria* L., but more recently henbane was suggested,[88] because of the symptoms and the availability of henbane in Nordic countries.[89] In some regions of France, it was believed that a silver cup would break if henbane was placed in it.[19]

Legend had it that to gain the love of a woman, a man would need to pick the henbane plant early in the morning naked whilst standing on one foot.[90] In Germany, henbane was burned to attract rain, although this also had the effect of sickening livestock, something that was usually put down to witches.[91]

In Ancient Greece, henbane had a strong association with death and was placed in tombs to make the dead 'forget their loved ones'. Burning henbane was thought by some to invoke restless souls and demons, but others used it to repel evil spirits.[76]

### 3.3.2   Hemlock

As well as being used in witches' brews, hemlock was thought to be the Devil's own property and snakes were thought to wriggle away from even a single leaf of the plant for fear of paralysis. It was also unlucky in the Christian tradition to have a festive tree made of hemlock (rather than fir),[54] possibly because of a legend that hemlock grew on the hill where Jesus was crucified.

### 3.3.3   Yew

Yew trees were sacred to the Druids, who considered them an emblem of immortality.[52] They were seen as powerfully magical and people often planted them alongside their houses for protection.[20,26] Yew was also dedicated to the gods of death[21,82] and was sacred to Hecate,[73] the Greek goddess associated with witchcraft, death, and necromancy (communicating with the dead). Like, blackthorn (see Chapter 10), yew was considered a good choice for magic wands,[52] and in the first[92] and last[93] Harry Potter books we learn that the wand of Tom Riddle, who becomes the darkest wizard who ever lived, is made of yew and phoenix feather.

## 3.4   MODERN MEDICAL USES

### 3.4.1   Henbane

Once the active ingredients were isolated from henbane in the 1820s, they could be used more safely, and hyoscyamine was sold as a prescription drug for indigestion and other gastrointestinal tract problems.[32,40] Henbane has now fallen out of prescribed use in the UK,[94] but as we saw in the previous chapter, another of its alkaloids, hyoscine (or scopolamine) is still used medicinally. Various parts of the plant are used in Traditional Chinese Medicine (TCM) including the root as an antimalarial, the leaves for abdominal pain and toothache, and ripe seeds to treat epilepsy.[95]

Although black henbane grows all over England, it is now most likely to be seen as a weed on dry sandy wasteland in the South.[37] Setting up your own business selling the plants is unlikely to be profitable however, as *Hyoscyamus niger* in the UK can only be sold in registered pharmacies under the supervision of a pharmacist.[19,96]

### 3.4.2 Hemlock

Hemlock is now medically obsolete and strongly advised against as an internal drug because of the danger of poisoning and lack of evidence of therapeutic benefits.[51] There is no legally permitted dose of the leaf or fruit unless made available by a prescription from a registered doctor or dentist.[96] However, some homeopathic dilutions and ointments made from flowering hemlock are still used externally in rubs or ointments.[32] It is recommended by homeopathy handbooks to treat dizziness, exhaustion and menstruation problems.[97]

### 3.4.3 Yew

Unlike many of the poisonous plants we will meet in this book, yew is sold openly in garden centres to be used in hedges (as we heard were planted at Malfoy Manor in *Harry Potter and the Deathly Hallows*), and even Christmas wreaths, although some websites do come with warnings that it is poisonous. It might seem odd that something so toxic is so popular, but yew's small leaves help it to form into precise shapes when trimmed.[27]

Yew is also used in homeopathic tinctures for poor digestion and skin pustules.[32] Taxanes originally extracted from the bark of yew trees are very promising chemotherapy agents *e.g.*, **docetaxel** or paclitaxel (sold with the name Yewtaxan®).[98] These drugs are now more likely to be synthetic (made in the lab from scratch) or semisynthetic (made in the lab from starting materials found in yew plants).[99,100]

## 3.5 MODERN TOXICOLOGY CASES

### 3.5.1 Henbane

Because of the unpleasant odour and taste of henbane, accidental poisonings with the plant are fortunately rare, however its roots have occasionally been gathered and eaten.[30] Accidental poisonings with henbane are most likely due to confusion with other plants, such as viper's grass (*Scorzonera hispanica*), wild parsnip (*Pastinaca sativa*), chicory, celery or poppy seeds.[21,37,57,58]

Henbane leaves have been made into tea since Roman times,[41] and in 2006, a 71-year-old man in Turkey ended up in A&E after drinking henbane tea to treat his asthma. His vision was distorted and his mouth became dry about 2 h after drinking it.[101]

A 65-year-old woman in Canada ingested the cooked root of an "odd parsnip" from her garden in 2017. The root of the hemlock plant is white and fleshy similar to parsnips,[81] so this is a common mistake. In A&E she was agitated and hallucinating with dilated pupils. She was discharged after three days but continued to suffer fatigue for another three weeks. Her 71-year-old husband also ate the plants and was so agitated he had to be restrained for 16 h. He was discharged after two days, but months later still had no memory of the unfortunate events.[31]

Treatment for henbane poisoning, if discovered before the plant has been fully digested, includes gastric lavage (pumping out and washing the stomach) and temperature-lowering measures with wet cloths. The antipsychotic drug **chlorpromazine** can also be given for severe excitation.[32] The antidote **physostigmine** we met in the previous chapter, can also be used to reverse henbane poisoning.

Another danger is adverse drug interactions – where herbal medicines work together with prescription or over-the-counter medications such as antidepressants and antihistamines[32] to produce negative effects. For example, henbane can exacerbate the sleepiness caused by the antihistamine diphenhydramine.[102]

Teenagers have also been known to use henbane as a recreational substance in several different countries, either by chewing the leaves or drinking an infusion of the seeds.[38] The consequences of this can be serious, and in 2014 in Belarus, 10 teenagers were taken to hospital with poisoning caused by henbane, Datura (Chapter 7) or castor bean plants (containing ricin).[103] In Israel, there is a more recent trend for chewing henbane flowers in the hope of producing euphoria,[53] although cases of this were also reported in Australia in the 1970s.[104]

### 3.5.2 Hemlock

In Italy, 17 poisonings were caused between 1972 and 1990 by people eating birds that had survived eating hemlock seeds or buds earlier in the Spring.[51] Only 13 of the patients survived,[62] and many developed rhabdomyolysis (the breakdown of dam-aged skeletal muscle and the release of myoglobin into the bloodstream) which can cause kidney damage and even kidney failure.[69] It can be treated by diuresis (maintaining adequate urine output by giving the patient a large volume of water) and changing the pH of the urine to make it more alkaline.[69]

One of the curious things about poisonous plants is that they don't affect all species in the same way – just because a cater-pillar, bird or squirrel ate a plant with no problems, it isn't safe to assume we can do the same.[55] This is sometimes because the animals eat the plants earlier in the season when they are less potent, or because the animals are able to tolerate or detoxify the poisons,[105] or because they lack the human biological systems that the poisons act on.

Large mammals seem to be less resistant than birds to hem-lock poisoning, and it sometimes occurs in horses, cattle, sheep and pigs in Europe.[59,106] These poisonings happen when hem-lock plants accidentally end up in hay or silage, or are foraged by the animals in meadows and grass pastures.[65] The symptoms in animals are similar to those in humans, with frequent urination, birth defects, trembling and staggering the most noticeable to the owners.[107]

In 2011, a 59-year-old farmer in Turkey was sent to A&E after eating some parsley-like plants he had picked after watching a TV show about herbal medicine.[69] For those who accidentally over-dose on hemlock, there is no specific antidote so the best treat-ment, as was used in this case, is activated charcoal.[51,69] Activated charcoal is a kind of carbon that has a large surface area and is able to bind to other substances such as drugs or poisons. This binding prevents them from being absorbed from the stomach and into the blood. Patients drink 50–100 g of charcoal powder mixed with water. The Turkish farmer was also given **atropine** to treat his heart symptoms, and recovered fully after two days.[69]

In an unusual case from the USA in 2017, a 30-year-old man was taken to A&E after suffering a cardiac arrest (heart attack) in an apparent suicide attempt.[106] His note indicated he had injected himself with hemlock, and a needle and syringe containing an unknown liquid were found at the scene. The liquid was tested using **LC-MS** and found to contain coniine, *N*-methylconiine, and **conhydrine**.[106] He recovered after 23 days in hospital.[106]

In another apparent suicide attempt, a 28-year-old man in the USA in 2013 was found unconscious with a jar containing a mix of coniine, amitriptyline, and **diazepam** (Valium). He remained in a coma in hospital for 14 days.[108] Ironically, diazepam is often used to treat the anxiety, restlessness and convulsions associated with hemlock poisoning.[30,69]

A couple were admitted to A&E in 2009 in Italy after eating wild plants they had gathered from a ditch that they thought were fennel. Their symptoms included nausea, painful leg cramps, a rash, dilated and unresponsive pupils and dizziness, and the plants were identified as hemlock. Both survived and were discharged from hospital after three days.[55]

Poisonings in children also happen, and in New Zealand between 2003 and 2010 the Poison Centre received 77 calls about hemlock poisoning, half of which concerned children.[67] A 6-year-old girl in Turkey was taken to A&E in 2016 after eating what she thought was parsley and complaining of a burning in her mouth and excessive salivation. She also suffered from prolonged coagulation, meaning that her blood was taking too long to clot, the opposite of Hamlet Snr's "curd" symptoms. She was discharged recovered after three days.[109] In the USA in 1995, a 4-year-old boy was admitted unconscious to A&E after eating the tops of what he thought were wild carrots (another name for hemlock is "carrot fern"[65]). Fortunately, the boy had vomited up much of the plant material already and was discharged after two days.[110] The leaves were analysed by **gas chromatography-mass spectrometry** (GC-MS) and found to contain gamma-coniceine. In 2009 in the USA, a 2-year-old boy was hospitalised after his 4-year-old sister gave him a green plant to eat. Fortunately some of the plant remained to be identified as hemlock, and the boy recovered by the evening.[60] In a more serious case, a 3-year-old ate over 140 g of hemlock leaves that were growing in a backyard in Australia in 1995 and died within a few hours.[111]

GC-MS is another "hyphenated" technique, joining two machines together. The first one, the gas chromatograph, is a separation technique where the sample (in this case a plant extract in a solvent) is heated to a high temperature (190 °C) to turn it into a gas. The gas is then moved down a long (30 m) and thin (0.25 µm) metal tube or "column" by an unreactive "carrier gas" such as helium. By the time the gas reaches the end of the column, most of the chemicals have spread out and emerge one at a time to be analysed by the mass spectrometer (see Chapter 2). Occasionally chemicals with very similar structures, including enantiomers (mirror images), will emerge together, or "coelute", so we cannot tell them apart. In these instances special columns may be used, or a chemical modification known as **derivatisation** (see Chapter 10) is needed. GC-MS works well for many drugs and poisons, but the heat destroys others such as amphetamines, so **LC-MS** (see Chapter 2) may be used instead.

### 3.5.3 Yew

Consumption by accident is relatively frequent, especially of the red berries by children up to the age of three (although these are not poisonous).[82,112] In one case in Germany in 2003 a patient with schizophrenia died after drinking a cup of yew tea as an alternative to their prescribed antipsychotic medication.[112] In Canada in 2021, a 40-year-old woman died on a video call after accidentally picking and eating yew leaves; she had 180 g of the plant in her stomach at the autopsy.[113]

Suicide attempts with yew needles also happen.[59] In 2014 in the Czech Republic a 25-year-old male patient was admitted to A&E with acidosis (body fluids containing too much acid) after eating yew needles, and became unconscious and died.[82] A 44-year-old male farmer was admitted to hospitalised in 2010 in Italy with vomiting and loss of consciousness in an apparent suicide attempt with yew, although at first doctors thought he was having a heart attack.[114] In Austria between 2007 and 2017 there were 16 attempted suicides with yew; in one fatality a 35-year-old woman was found dead after eating yew needles, in another a 20-year-old woman died after drinking yew extract.[115] Closer to home in the UK, the National Poisons Information Service recorded 443 cases of exposure to yew, nine of which were by accidentally inhaling the fumes when sawing or burning yew wood.[116]

Although the majority of cases involve self-harm, occasionally yew poisonings are due to attempts to get high from the plant, and there were five cases of this between 1995 and 2009 in Switzerland.[117]

Although the alkaloids in yew are well known, toxicologists often look for the chemical **3,5-dimethoxyphenol (3,5-DMP)** instead,[118] which is the main metabolite (body breakdown product) of taxicatine in the blood. In one case, a 46-year-old man was brought to A&E in Poland in 2016 with nausea, vomiting and cardiac arrest after eating a yew leaf sandwich. His serum was positive for 3,5-DMP but he survived and was eventually discharged.[119] Finding 3,5-DMP alone in the body is not definite proof of yew poisoning however, as other plants such as *Rosa chinensis*, *Ruta graveolens* or grapes could be the source.[118]

This approach of looking for metabolites or markers rather than the parent compound is used quite widely by toxicologists. Either because the parent chemical is so short-lived that looking for it is a waste of time (*e.g.*, **heroin**), or because the body activates an inert drug by turning it into something else (known as a "pro-drug"). The most common example of this is **aspirin**, which is transformed into **salicylate** by the liver. It is the salicylate that is responsible for reducing pain and inflammation, and this is usually measured in body fluids using **high-performance liquid chromatography with diode array detection** (**HPLC-DAD**, see Chapter 7). Although aspirin is widely available now, the bark of the willow tree (*Salix alba*, Figure 3.6) was our original source of salicylates.

Yew is different to other plants, in that its needles are often found in the stomachs of poisoning victims because they are so slow to digest (a layer on the yew needle surface prevents the alkaloids from being released). In five cases of yew poisoning in Germany in 2006, four victims in their twenties and one teenager were all found dead and had undigested yew leaves in their stomachs.[120] Forensic botanists or forensic plant scientists can examine bits of plant material found in the stomach, in vomit[121] or even at the scene.[113] As well as identifying mystery material, they can help determine time of death, and connect people to crime scenes through traces of evidence on clothing or shoes.[122]

Toxicologists prefer to analyse blood over stomach contents because the latter requires a lot of clean-up. Although stomach contents can hold some very useful clues, it is only a qualitative strategy, as the concentration of a drug or poison in the stomach contents isn't really meaningful; if we find a drug or poison in the stomach it tells the toxicologist that it was eaten (knowingly or otherwise) by the deceased. But if it's still in the stomach, it isn't having an effect on the body, as only drugs that have

**Figure 3.6**    A willow tree, taken at The Poison Garden in Alnwick.

passed into the blood can cause symptoms. Depending on how quickly the person died after eating the poison, most of it could still be in the stomach. So a high concentration in the stomach, doesn't necessarily mean a high concentration in the blood. Low concentrations in the stomach can also be misleading as drugs can react with the stomach acid and break down before they are absorbed.[123] After death, drugs can also travel across the lining of the stomach and move around the body in a process known as "redistribution".[120] Sometimes whole undissolved tablets are found in the stomach and compared to databases of photos of pills, so we know what to target in the blood.

## 3.6    CATCHING THE CHEMICAL CULPRIT

In the case of Hamlet Snr, there was no external evidence to suggest that a crime had been committed. Even if the ears were examined in detail, the weapon was probably a brownish oily extract that looked very similar to earwax.[6] This problem persists today, as the autopsies of poisoning victims often reveal few clues (or are "unremarkable" in Pathologists' terms) as to the cause of death.[120]

Sometimes remains of the plants are found at the scene of a poisoning (if not tidied away by relatives) and these can be analysed[55] to give vital hints to the toxicologist or A&E doctor. If the patient has eaten the lot, sometimes a photograph is passed to a botanist for identification, but these pictures can lack the unique characteristics of some poisonous plants. This happened in a case in 2020 of a 53-year-old woman in the USA who consumed what she thought was sweet fennel. Although her symptoms were consistent with coniine poisoning, three botany experts were convinced from the photographs that the plant was not hemlock. Only after new plant samples were collected from the foraging spot, did they resemble hemlock.[56] Another approach is to extract the poisonous chemicals from the plants as crystals and look at them using a microscope,[55] as many substances make distinctive and beautiful crystals in these quick tests.[124]

## 3.7 CASE CLOSED

Although we can't be sure of the identity of our mystery poison hebenon, the three plants henbane, hemlock and yew are all candidates. We also don't know how much poison was used on the king, but Shakespeare tells us it was a fatal dose.[69] We saw from our real-life case involving hemlock that it causes blood to coagulate more slowly, whereas experiments have shown that henbane quickens blood coagulation,[12] making this the most likely culprit. As with many toxic encounters in fiction, the poison is there to represent something wider than just a painful death, and in *Hamlet* it becomes an all-embracing metaphor of the corruption and decay of the Danish Court.[125]

There were another two poisonings in *Hamlet*, where the poison itself was not named – one is dabbed on the end of a sword to be used in a duel between Hamlet Jnr and Laertes, and the other added to a cup, which Queen Gertrude drinks. These could have been cyanide (Chapter 10), **aconitine** (Chapter 5) or even the more exotic toxin curare, an arrow poison from the Amazon, which was a mixture of the *Strychnos* (Chapter 8), *Erythrina* and *Chondrodendron* plants.[2,126]

## REFERENCES

1. W. Shakespeare, *Hamlet*, Oxford University Press, Oxford, 1998.
2. K. Harkup, *Death by Shakespeare: Snakebites, Stabbings and Broken Hearts*, Bloomsbury, London, 2020.
3. W. Turner, *A New Herball*, Cambridge University Press, Cambridge, 1989.
4. E. Tabor, *Econ. Bot.*, 1970, **24**, 81–94.
5. Anonymous, *Herbal*, Rycharde Banckes, London, 1525.
6. R. R. Simpson, *J. Laryngol. Otol.*, 2007, **64**, 342–352.
7. M. Chamberlin, *Old Wives' Tales: The History of Remedies, Charms and Spells*, The History Press, Cheltenham, 2020.
8. A. S. Harper-Leatherman and J. R. Miecznikowski, *J. Chem. Educ.*, 2012, **89**, 629–635.
9. J. P. André, *J. Chem. Educ.*, 2013, **90**, 352–357.
10. R. J. Huxtable, *Perspect. Biol. Med.*, 1993, **36**, 262–280.
11. H. Bradley, *Mod. Lang. Rev.*, 1920, **15**, 85–87.
12. D. I. Macht, *Bull. Hist. Med.*, 1949, **23**, 186–194.
13. *The New Oxford Annotated Bible (New Revised Standard Version)*, ed. M. D. Coogan, M. Z. Brettler, C. A. Newsom and P. Perkins, Oxford University Press, Oxford, 2006.
14. C. Marlowe, *The Jew of Malta*, Cambridge University Press, Cambridge, 1973.
15. B. Jonson, *The Masque of Queens*, Yale University Press, New Haven, CT, 1969.
16. Apollonius Rhodius, *Argonautica*, Harvard University Press, Cambridge, MA, 2009.
17. S. Brown, in *The Iris: Behind the Scenes at the Getty*, 2015, vol. 2020.
18. The Powerful Solanaceae: Henbane, https://www.fs.fed.us/wildflowers/ethnobotany/Mind_and_Spirit/henbane.shtml, accessed June 2020.
19. K. Fatur, *Econ. Bot.*, 2020, **20**, 1–19.
20. G. Grigson, *The Englishman's Flora*, Helicon, Herzelia, 1996.
21. D. Frohne and H. J. Pfänder, *Poisonous Plants: A Handbook for Doctors, Pharmacists, Toxicologists, Biologists and Veterinarians*, Manson Publishing, London, 2nd edn, 2004.
22. A. Börsch-Haubold, *Sci. Sch.*, 2007, **4**, 50–55.
23. M. Blamey and C. Grey-Wilson, *Mediterranean Wild Flowers*, Harper Collins, London, 1993.

24. *Gerard's Herbal*, ed. M. Woodward, Senate, London, 1994.

25. L. S. Nelson, M. A. Howland, N. A. Lewin, S. W. Smith, L. R. Goldfrank and R. S. Hoffman, *Goldfrank's Toxicologic Emergencies*, 11th edn, McGraw Hill, New York, 2018.

26. S. Lawrence, *Witch's Garden: Plants in Folklore, Magic and Traditional Medicine*, Welbeck, London, 2020.

27. R. Bevan-Jones, *Poisonous Plants: A Cultural and Social History*, Windgather Press, Oxford, 2009.

28. M. R. Cooper, A. W. Johnson and E. A. Dauncey, *Poisonous Plants and Fungi: An Illustrated Guide*, The Stationery Office, Norwich, 2nd edn, 2003.

29. M. R. Lee, *J. R. Coll. Physicians Edinburgh*, 2006, **36**, 366–373.

30. A. Alizadeh, M. Moshiri, J. Alizadeh and M. Balali-Mood, *Avicenna J. Phytomed.*, 2014, **4**, 297–311.

31. T. A. Shams, S. Gosselin and R. Chuang, *Toxicol. Commun.*, 2017, **1**, 37–40.

32. Medical Economics Company, *PDR for Herbal Medicines*, Medical Economics Company, Montvale, NJ, 1998.

33. T. Hargreaves, *Poisons and Poisonings: Death by Stealth*, RSC Publishing, Cambridge, 2017.

34. A. J. Carter, *Br. Med. J.*, 1996, **313**, 1630–1632.

35. P. Dioscorides, *The Greek Herbal of Dioscorides*, Haner Publishing, New York, 1959.

36. M. Grieve, *A Modern Herbal*, Dover Publications, New York, 2nd edn, 1971.

37. P. M. North, *Poisonous Plants and Fungi*, Blandford Press, London, 1967.

38. K. Fatur and S. Kreft, *Toxicon*, 2020, **177**, 52–88.

39. P. Wexler, *History of Toxicology and Environmental Health: Toxicology in Antiquity*, Academic Press, Amsterdam, 2014.

40. J. M. Riddle, *Goddesses, Elixirs, and Witches*, Palgrave MacMillan, New York, 2010.

41. J. Mann, *Murder Magic and Medicine*, Oxford University Press, Oxford, 1992.

42. W. Shakespeare, *Macbeth*, Cambridge University Press, Cambridge, 1960.

43. M. Brown, *Death in the Garden: Poisonous Plants & their Use Throughout History*, Pen & Sword Books Ltd, Barnsley, 2018.

44. J. Gerard and T. Johnson, *The Herball or Generall Historie of Plantes*, 2nd edn, Adam Islip, Joice Norton & Richard Whitakers, London, 1636.

45. V. Thomas and N. Faircloth, *Shakespeare's Plants and Gardens: A Dictionary*, Bloomsbury, London, 2016.
46. W. Shakespeare, *The Tragedy of King Lear*, 2nd edn, Cambridge University Press, Cambridge, 2005.
47. *A New Grove Dictionary of Opera*, ed. S. Sadie, Oxford University Press, Oxford, 1997.
48. *Giacomo Puccini Suor Angelica: Complete Libretto with Music Highlight Examples*, ed. B. D. Fisher, Opera Journeys Publishing, Coral Gables, 2007.
49. J. Keats, *John Keats the Major Works*, Oxford University Press, Oxford, 2001.
50. B. Hubbard, *Poison: The History of Potions, Powders and Murderous Practitioners*, Welbeck, London, 2020.
51. J. Emsley, *More Molecules of Murder*, RSC Publishing, Cambridge, 2017.
52. A. Chevallier, *Encyclopedia of Medicinal Plants*, DK Publishing, St Leonards, 2001.
53. Y. Gaillard and G. Pepin, *J. Chromatogr. B: Biomed. Sci. Appl.*, 1999, **733**, 181–229.
54. C. M. Skinner, Myths and Legends of Flowers, Trees, Fruits, and *Plants: In all Ages and in all Climes*, J. B. Lippincott, Philadephia, 1911.
55. M. L. Colombo, K. Marangon, C. Locatelli, M. Giacchè, R. Zulli and P. Restani, *J. Pharm. Sci. Res.*, 2009, **1**, 43–47.
56. J. M. Rague, L. S. Halmo and K. Heard, *American College of Medical Toxicology 2020 Annual Scientific Meeting New York*, 2020, p. 163.
57. H.-Y. Chen, H. Horng, F. Rowley and C. Smollin, *Clin. Toxicol.*, 2017, **55**, 155–156.
58. D. G. Spoerke, A. H. Hall, C. D. Dodson, F. R. Stermitz, C. H. Swanson and B. H. Rumack, *J. Emerg. Med.*, 1987, **5**, 385–388.
59. D. Frohne and H. J. Pfänder, *A Colour Atlas of Poisonous Plants*, Wolfe Publishing Ltd, London, 1983.
60. P. L. West, B. Z. Horowitz, M. T. Montanaro and J. N. Lindsay, *Pediatr. Emerg. Care*, 2009, **25**, 761–763.
61. G. Ponting, *Shakespeare's Fantastic Garlands*, Millers Dale, Eastleigh, 2008.
62. D. Rizzi, C. Basile, A. Di Maggio, A. Sebastio, F. Introna Jr, R. Rizzi, A. Scatizzi, S. De Marco and J. E. Smialek, *Nephrol., Dial., Transplant.*, 1991, **6**, 939–943.

63. M. Levine, A.-M. Ruha, K. Graeme, D. E. Brooks, J. Canning and S. C. Curry, *Chest*, 2011, **140**, 1357–1370.
64. R. Vickery, *Garlands, Conckers and Mother-Die: British and Irish Plant-Lore*, Continuum, London, 2010.
65. H. Hotti and H. Rischer, *Molecules*, 2017, **22**, 1962.
66. T. Reynolds, *Phytochemistry*, 2005, **66**, 1399–1406.
67. R. J. Slaughter, M. G. Beasley, B. S. Lambie, G. T. Wilkins and L. J. Schep, *N. Z. Med. J.*, 2012, **125**, 87–118.
68. S. Funayama and G. A. Cordell, *Alkaloids: A Treasury of Poisons and Medicines*, Elsevier, Amsterdam, 2015.
69. A. Erenler, A. Baydin, L. Duran, T. Yardan and B. Turkoz, *Hong Kong J. Emerg. Med.*, 2011, **18**, 235–238.
70. Anonymous, *The Poison Garden*, The Alnwick Garden, Alnwick, 2005.
71. J. Timbrell, *The Poison Paradox: Chemicals as Friends and Foes*, Oxford University Press, Oxford, 2005.
72. W. Shakespeare, *History of Richard II*, Cambridge University Press, Cambridge, 2nd edn, 2003.
73. A. Major, *Scholia*, 1995, **4**, 101–104.
74. W. Shakespeare, *Romeo and Juliet*, Cambridge University Press, Cambridge, 2003.
75. A. Stewart, *Wicked Plants*, Algonquin Books of Chapel Hill, Chapel Hill, NC, 2009.
76. F. Inkwright, *Botanical Curses and Poisons*, Liminal 11, London, 2021.
77. J. K. Rowling, *Harry Potter and the Goblet of Fire*, Bloomsbury, London, 2000.
78. A. Tennyson, *Tennyson: In Memoriam*, Clarendon Press, Oxford, 1984.
79. W. Shakespeare, *Titus Andronicus*, Cambridge University Press, Cambridge, 2006.
80. W. Shakespeare, *Twelfth Night or What You Will*, Cambridge University Press, Cambridge, 2003.
81. J. Beyer, O. H. Drummer and H. H. Maurer, *Forensic Sci. Int.*, 2009, **185**, 1–9.
82. O. Piskač, J. Stříbrný, H. Rakovcová and M. Malý, *Cor Vasa*, 2015, **57**, e234–e238.
83. S. Schultes, A. Hofmann and C. Rätsch, *Plants of the Gods*, Healing Arts Press, Rochester, VT, 2001.

84. R. Richardson, *Britain's Wild Flowers: A Treasury of Traditions, Superstitions, Remedies and Literature*, The National Trust, London, 2017.
85. M. R. Lee, *J. R. Coll. Physicians Edinburgh*, 2006, **36**, 278–285.
86. E. A. Campbell, *Vic. Lit. Cult.*, 2007, **35**, 607–615.
87. J. C. Bucknill, *The Medical Knowledge of Shakespeare*, Longman & Co, London, 1860.
88. K. Fatur, *J. Ethnopharmacol.*, 2019, **244**, 112151.
89. T. Lempiäinen, *Ann. Bot. Fenn.*, 1991, **28**, 261–272.
90. S. T. Dietz, *The Complete Language of Flowers: A Definitive and Illustrated History*, Wellfleet Press, New York, 2020.
91. F. Inkwright, *Folk Magic and Healing: An Unusual History of Everyday Plants*, Liminal 11, London, 2019.
92. J. K. Rowling, *Harry Potter and the Philosopher's Stone*, Bloomsbury, London, 1997.
93. J. K. Rowling, *Harry Potter and the Deathly Hallows*, Bloomsbury, London, 2007.
94. Royal Pharmaceutical Society, *British National Formulary*, BNF Publications, London, 2021.
95. J. Zhou, G. Xie and X. Yan, *Encyclopedia of Traditional Chinese Medicines*, Springer, Heidelberg, 2011.
96. Banned and restricted herbal ingredients, https://www.gov.uk/government/publications/list-of-banned-or-restricted-herbal-ingredients-for-medicinal-use/banned-and-restricted-herbal-ingredients, accessed May 2020.
97. M. Castro, *The Complete Homeopathy Handbook: A Guide to Everyday Health Care*, Macmillan, London, 1990.
98. *Clarke's Analysis of Drugs and Poisons*, ed. A. C. Moffatt, D. Osselton and B. Widdop, Pharmaceutical Press, London, 4th edn, 2011.
99. P. A. Thomas and A. Polwart, *J. Ecol.*, 2003, **91**, 489–524.
100. A. Foster, *The Medicinal Plant Collection at the University of Oxford Botanic Garden*, Wellcome Trust, Oxford, 2010.
101. H. Erkal, Y. Ozyurt and Z. Arikan, *Turk. J. Genet.*, 2006, **9**, 188–191.
102. H. H. Tsai, H. W. Lin, A. Simon Pickard, H. Y. Tsai and G. B. Mahady, *Int. J. Clin. Pract.*, 2012, **66**, 1056–1078.
103. V. Lelevich, H. Vinitskaya, Y. Sarana and E. Tischenko, *Cent. Eur. J. Sport Sci. Med.*, 2016, **15**, 85–94.

104. J. M. Sands and R. Sands, *Med. J. Aust.*, 1976, **2**, 55–58.
105. S. R. Whitehead and M. D. Bowers, *Am. Nat.*, 2013, **182**, 563–577.
106. D. Brtalik, J. Stopyra and J. Hannum, *J. Med. Toxicol.*, 2017, **13**, 180–182.
107. C. Cortinovis and F. Caloni, *Toxins*, 2015, **7**, 5301–5307.
108. D. D. Lung, B. J. Scott, A. H. B. Wu and R. R. Gerona, *Muscle Nerve*, 2013, **48**, 823–827.
109. C. Konca, Z. Kahramaner, M. Bosnak and H. Kocamaz, *Turk. J. Emerg. Med.*, 2016, **14**, 34–36.
110. B. S. Frank, W. B. Michelson, K. E. Panter and D. R. Gardner, *West J. Med.*, 1995, **163**, 573–574.
111. O. H. Drummer, K. L. Crump, M. H. Phelan, A. N. Roberts and P. J. Bedford, *Med. J. Aust.*, 1995, **162**, 592–593.
112. J. Beike, B. Karger, T. Meiners, B. Brinkmann and H. Köhler, *Int. J. Legal Med.*, 2003, **117**, 335–339.
113. E. W. L. Brooks-Lim, S. A. Mérette, B. J. Hawkins, C. Maxwell, A. Washbrook and A. M. Shapiro, *J. Forensic Sci.*, 2022, **67**, 820–826.
114. C. Panzeri, G. Bacis, F. Ferri, G. Rinaldi, A. Persico, F. Uberti and P. Restani, *Clin. Toxicol.*, 2010, **48**, 463–465.
115. S. Dorner-Schulmeister, K. Bartecka-Mino and A. Holzer, *40th International Congress of the European Association of Poisons Centres and Clinical Toxicologists (EAPCCT)*, Tallinn, Estonia, 2020, p. 537.
116. R. M. Gowda, R. A. Cohen and I. A. Khan, *Heart*, 2003, **89**, e14.
117. J. Fuchs, C. Rauber-Lüthy, H. Kupferschmidt, J. Kupper, G.-A. Kullak-Ublick and A. Ceschi, *Clin. Toxicol.*, 2011, **49**, 671–680.
118. F. Musshoff and B. Madea, *Int. J. Legal Med.*, 2008, **122**, 357–358.
119. M. Kobusiak-Prokopowicz, A. Marciniak, S. Ślusarczyk, K. Ściborski, A. Stachurska, A. Mysiak and A. Matkowski, *BMC Pharmacol. Toxicol.*, 2016, **17**, 41.
120. J. Pietsch, K. Schulz, U. Schmidt, H. Andresen, B. Schwarze and J. Dreβler, *Int. J. Legal Med.*, 2007, **121**, 417–422.
121. F. Veit, M. Gürler, A. Nebel, C. G. Birngruber, R. B. Dettmeyer and W. Martz, *Forensic Sci. Int. Rep.*, 2020, **2**, 100158.

122. J. H. Bock and D. O. Norris, in *Forensic Plant Science*, ed. J. H. Bock and D. O. Norris, Academic Press, San Diego, 2016, ch. 1, pp. 1–22.
123. L. Frommherz, P. Kintz, H. Kijewski, H. Köhler, M. Lehr, B. Brinkmann and J. Beike, *Int. J. Legal Med.*, 2006, **120**, 346–351.
124. L. Elie, M. Baron, R. Croxton and M. Elie, *Forensic Sci. Int.*, 2012, **214**, 182–188.
125. P. Sadowski, in *British Shakespeare Association Conference*, Hull, 2016, pp. 1–8.
126. T. Stone and G. Darlington, *Pills, Potions, Poisons: How Drugs Work*, Oxford University Press, Oxford, 2000.

CHAPTER 4

# The Flattering Truth of Sleep

If you see a term that's **bold** it's defined in the Glossary. Only the first time that the word appears in the chapter will it be indicated in this way.

---

**Case History: Murder–Suicide**

*A teenage girl is found dead and laid to rest in the family tomb. Investigations reveal that she has drunk a potion given to her by a Friar. Twenty-seven hours later she comes back to life to discover her lover Romeo dead by her side, then takes her own life with a dagger.*

FRIAR:      ...Take thou this vial; being then in bed,
And this distilling liquor drink thou off;
When presently through all thy veins shall run
A cold a drowsy humor; for no pulse
Shall keep his native progress, but surcease;
No warmth, no þreath, shall testify thou livest;
The roses in thy lips and cheeks shall fade
To wanny ashes, thy eyes' windows fall
Like death when he shuts up the day of life;
Each part, deprived of supple government,

---

Poisonous Tales: A Forensic Examination of Poisons in Fiction
By Hilary Hamnett
© Hilary Hamnett 2023
Published by the Royal Society of Chemistry, www.rsc.org

Shall, stiff and stark and cold, appear like death
Thou shalt continue two-and-fourty hours,
And then awake as from a pleasant sleep.

Act IV, scene i

*The Tragedy of Romeo & Juliet* by William Shakespeare,[1] 1594

**Figure 4.1**  Romeo and Juliet last scene, by J. Northcode.

## 4.1  THE INVESTIGATION

We have seen already that a potion made of mandrake and poppy can give a convincing appearance of death (see Chapter 2). In our previous case study the symptoms were not described, but in *Romeo & Juliet* (Figure 4.1), the Friar gives an account to Juliet before she drinks the "liquor" (sleeping draught) saying that she will feel drowsy, and appear pale and stiff like death for 42 h. Her aim is to avoid an arranged marriage to Paris.[2] When Juliet is initially discovered in her bedroom, she is described as cold with settled blood and stiff joints. Her coma is convincing enough for her relatives to place her in the Capulet family tomb, news of which eventually reaches Romeo.

Later on she is found in the tomb with fatal knife wounds alongside the also-poisoned Romeo, a case that in modern times, would probably be initially recorded as a "murder–suicide", with the suspicion that one of the couple had killed the other, then taken their own life. Of course post-mortem toxicology for both lovers would have revealed a different story, and we will discuss Romeo's potion in Chapter 5. Interestingly, before he finds out about Juliet's 'death', Romeo dreams about his own demise and Juliet bringing him back to life with a kiss; the "flattering truth of sleep".

The Friar does not tell us the ingredients in Juliet's potion but it is now thought to have been deadly nightshade,[3-6] because its poisonous properties were described by Gerard at the time. He tells us that deadly nightshade is so furious and deadly that it causes sleep, troubles the mind, and brings on madness. He implores us to banish it from our gardens in case any children come across it.[7] Today it might seem odd that a Friar (priest) would be providing a poisonous potion, but monasteries often had medicinal herb gardens for mixing up natural remedies for those within and outside the order,[4,8] and religious men were some of the most highly educated at the time.[6] During this period, knowledge came from and was kept within religious orders.[9]

## 4.2   THE PLANT BEHIND THE STORY

Deadly nightshade (*Atropa belladonna*) also known as "bane-wort", "maddening herb", or "sleepy nightshade" has a long history as a classic poison.[10,11] The plant has single brownish violet bell-shaped flowers and glossy black fruit known as "Devil's berries" (Figure 4.2).[12,13] The leaves are always paired – a small and large leaf together. The flowers appear from June to August,[14] and deadly nightshade can be seen on the edge of woods and in sparse forests, particularly in the South of England.[12,13]

All parts of the deadly nightshade plant contain the mix of alkaloids **scopolamine, hyoscyamine** and **atropine**.[15] These chemicals are sometimes called the "active principles" of the plant.[16] The flowers, seeds and green (unripe) fruits contain mainly **hyoscyamine**,[17] but as the fruits ripen and turn red then black, it converts to **atropine**. This combination of chemicals is similar to the mandrakes we met in Chapter 2, and this is because the two plants are in the same family, Solanaceae (which also includes tomatoes and potatoes).[18] All of the plants in this family act on the **acetylcholine** levels in the body, because the two main alkaloids chemically resemble acetylcholine, although their potencies are not the same.[19] The roots of the deadly nightshade plant are the most toxic, followed by the stem, leaves, flowers and berries.[20]

**Figure 4.2**  (Left) The fruit of a deadly nightshade plant, taken at the Cambridge University Botanic Garden. (Right) The flowers of the deadly nightshade plant, taken at the Chelsea Physic Garden.

They are collected at different times of the year, with the leaves best gathered in the summer and the roots in the autumn.[21] All of the alkaloids withstand drying and boiling,[14] and during storage of the plant, **apoatropine** can be formed.[22] Belladonna has a strong unpleasant smell, and a sharp and bitter taste.[22]

Squeezing the juice of deadly nightshade berries into the eyes makes the pupils dilate, producing a 'doe-eyed' beauty, and **atropine** is the chemical responsible for this. As well as its cosmetic effect, atropine also allows the retina to be seen clearly and is still used today by specialist eye doctors (in 2020 in England nearly 55 000 prescriptions for atropine eye drops were filled[23]).[3,11,24] Atropine can paralyse the muscles of the pupils when dilated, blurring the vision of the patient (described in medical textbooks as 'blind as a bat'),[11] and eye hospital patients are often advised not to drive home.[25]

Deadly nightshade previously had some other uses in medicine in plasters (known as "belladonna bandages"), liniments (oily liquids) and ointments.[14] But use of these plasters is associated with dermatitis, blurred vision and dizziness.[26] One particular brand of belladonna plasters also contained monkshood (Chapter 5), and the poisons could be extracted from the plasters for use as a murder weapon.[11] This extraction of drugs for nefarious purposes is still seen today from patches containing the powerful painkiller fentanyl.[27] The powdered root or leaves of deadly nightshade have been used in the past internally for liver and gall bladder complaints,[11,22] and tinctures were used for various complaints ranging from hay fever to haemorrhoids. These tinctures were made by mixing crushed plant material with alcohol and then pressing the mixture through a filter.[28] Other folk remedies included using belladonna for preventing miscarriage and for a 'difficult labour'.[29]

Moderate doses of belladonna give rise to symptoms such as overheating ('hot as a hare'), flushed skin ('red as a beet'), dry mouth ('dry as a bone') leading to extreme thirst, and excitement and delirium ('mad as a hen' or 'mad as a hatter').[11] Atropine poisoning victims also don't receive the body's usual signals to urinate ('full as a drum').[28] Symptoms generally begin within 1 h of ingestion,[30] and a high dose, such as the one Juliet drank, can bring on a coma lasting several hours.[31] In one case in Turkey in 2013, a 49-year-old woman fell into a coma for 24 h after eating a large number of deadly nightshade berries.[32] In France in 1996, three adults from a family of eight were hospitalized after cooking and eating belladonna berries they had mistaken for whortleberries (the children wisely declined the dish). One of the adults was in a coma for 30 h before recovering.[33]

### 4.3   POISONOUS PLOTS

We find the symptoms of belladonna poisoning in other fictional sources such as in Richard Wagner's opera *Tristan und Isolde* (Figure 4.3), which premiered in 1865. In the story, the two main characters drink a potion, which may have contained belladonna.

Tristan is the nephew of King Marke and princess Isolde is meant to be marrying the King. The pair try to form a suicide pact by drinking what they think is poison (a "draught of death"), but it turns out to have been swapped by Isolde's maid for a love potion.[34] Although we are not told the ingredients, within minutes they begin to feel their hearts racing, they are flushed, and their vision blurry, all symptoms that are consistent with belladonna intoxication[35,36]

**Figure 4.3**  Image from the Tristan und Isolde libretto, by Peter Hoffer. Reproduced from https://commons.wikimedia.org/wiki/File:Disegno_per_copertina_di_libretto,_disegno_di_Peter_Hoffer_per_Tristano_e_Isotta_(s.d.)_-_Archivio_Storico_Ricordi_ICON012467.jpg, under the terms of the CC BY-SA 4.0 license, https://creativecommons.org/licenses/by-sa/4.0/deed.en, attributed to Archivio Storico Ricordi.

*Sie [Isolde] trinkt. Dann wirft sie die Schale fort. Beide, von Schauder erfasst, blicken sich mit höchster Aufregung, doch mit starrer Haltung, unverwandt in die Augen, in deren Ausdruck der Todestrotz bald der Liebesglut weicht. Zittern ergreift sie. Sie fassen sich krampfhaft an das Herz und führen die Hand wieder an die Stirn. Dann suchen sie sich wieder mit dem Blick, senken ihn verwirrt und heften ihn wieder mit steigender Sehnsucht aufeinander*

ISOLDE:
*mit bebender Stimme*
Tristan!

*She [Isolde] drinks. Then she throws the shell away. Both of them, shuddering, look at each other with the utmost excitement, but with a rigid attitude, unblinking in the eyes, in whose expression the Devil soon gives way to the love-fire. Trembling seizes her. They seize the heart convulsively and put their hands back to their foreheads. Then they look again with the look, sink it confused and fix it again with increasing yearning*

ISOLDE:
*in a trembling voice*
Tristan!

| TRISTAN: | TRISTAN: |
|---|---|
| *überströmend* | *overflowing* |
| Isolde! | Isolde! |

| ISOLDE: | ISOLDE: |
|---|---|
| *an seine Brust sinkend* | *sinking to his chest* |
| Treuloser Holder! | Faithless Holder! |

| TRISTAN: | TRISTAN: |
|---|---|
| *mit Glut sie umfassend* | *with embers embracing them* |
| Seligste Frau! | Blessed woman! |
| *Sie verbleiben in stummer* | *They remain in mute embrace. From* |
| *Umarmung. Aus der Ferne* | *a distance you can hear trumpets* |
| *vernimmt man Trompeten* | |

Act I, scene v
*Tristan und Isolde* by Richard Wagner,[37] 1865

Later in the opera, Tristan is killed in a duel and Isolde starts to hallucinate and then dies. Although not mentioned in Shakespeare's play, the historical account of the real Macbeth involves poisoning an invading army with food and drink containing a relative of deadly nightshade.[38]

In John Keats' poem *Ode on Melancholy* (1820), on coping with sadness, various poisonous plants are mentioned including wolfsbane (Chapter 5), nightshade and yew (Chapter 3)

No, no, go not to Lethe, neither twist
Wolf's-bane, tight-rooted, for its poisonous wine;
Nor suffer thy pale forehead to be kiss'd
By nightshade, ruby grape of Proserpine;
Make not your rosary of yew-berries,
Nor let the beetle, nor the death-moth be°
Your mournful Psyche, nor the downy owl
A partner in your sorrow's mysteries;

*Ode on Melancholy* by John Keats,[39] 1820

The poet tells us not to take our own lives with nightshade berries, describing them as the "ruby grape of Prosperpine", who was a Roman goddess married to the ruler of the underworld (also known as Proserpina).[40]

However, Keats may have been confusing deadly nightshade berries, which are green or black, with the red fruit of the woody nightshade plant (*Solanum dulcamara*) also known as "bitter-sweet".[41] This plant (Figure 4.4) is also poisonous, but contains the alkaloid **solanine** rather than atropine. Solanine is responsible for poisonings with green potatoes, which are part of the same family.[42]

A folk story called *The Lyminster Knucker* features a poisonous Sussex churdle pie, which in one version is filled with deadly nightshade berries and fly agaric mushrooms and is fed to a terrifying water monster (the "knucker")

...there was a terrible churning, gurgling sound
and the knucker clutched his belly...
'Oh' he spluttered and keeled over, stone dead.

*The Lyminster Knucker* by Michael O'Leary,[43] 2016

The hero of the day, Jim Pulk, unfortunately meets a sticky end as he forgets to wash his hands after baking the pie, and then brushes his hands on his lips whilst drinking a celebratory pint in the village pub. In the story he also immediately falls down dead, which seems unlikely after such as small dose, and the time we have seen it takes for symptoms to come on. An image of the hero (before the pint) and the water monster can be found in a stained-glass window in Lyminster parish church. Another church, Furness Abbey in Cumbria was associated with belladonna by Wordsworth in his poem *The Prelude* (1805–6), where he describes the ruin as sitting in the Vale of Nightshade.[44]

The 1998 film *Practical Magic* features a belladonna poisoning. Sally Owens adds it to a bottle of tequila that her sister's abusive

**Figure 4.4**   The fruit of the woody nightshade plant.

partner is drinking. She plans just to knock him out, but over-
does it and he dies. When a policeman comes looking for him
a few days' later, he finds a bottle of deadly nightshade in their
greenhouse

*[Gary picks up a bottle and looks at the label]*

SALLY:        Belladonna. It's a sedative. People put it in their tea to
              relax, calm their nerves.
GARY:         Some people also use it as a poison.

*Practical Magic* screenplay by Alice Hoffman & Robin Swincord,[45] 1998

In the film, he suddenly passes out, but from the belladonna
symptoms we have seen, sedation is not one of them, so his sudden death is much more likely to be a cardiac arrest. In *The Hun-
ger Games* (2012), one of the competitors eats some "nightlock"
berries and dies. This is a made-up plant, probably a combination of nightshade and hemlock (Chapter 3) although the shiny
black berries are similar to those in Figure 4.2. In the film, others
'exit' the game after being stung by a nest of venomous wasps
(Chapter 9). The description of the poison berries killing a person 'within a minute' in the film is a bit far-fetched given what we
have seen happens in real poisonings.

The poem *The Poison Flower* (1882) by John Boyle O'Reilly is
thought be to about deadly nightshade.[46] Although belladonna
is not the only plant with purple hooded flowers or green leaves,
the other obvious candidate monkshood (see Chapter 5) has no
obvious smell

A rich pansy it was, with a small white lip
And a wonderful purple hood;
And your eye caught the sheen
Of its leaves, parrot-green,
Down the dim gothic aisles of the wood.
And its foliage rich on the moistureless sand
Made you long for its odorous breath;
But ah! 'twas to take
To your bosom a snake,
For its pestilent fragrance was death.

*The Poison Flower* by John Boyle O'Reilly,[47] 1882

There are many novels named after atropa belladonna, such as the melodrama *Bella Donna* by Robert Hitchens (1909), the anthology of weird fiction *Deadly Nightshade* by Peter Haining (1979), the murder mystery of the same title by Cynthia Riggs (2003), the fantasy novel *Belladonna* by Anne Bishop (2007), *Belladonna* by Daša Drndić (2017), the historical novel of the same title by Anbara Salam (2020) and, most recently, the historical thriller *Nightshade* by E S Thomson (2021). All of these books use the name to signify a witch or beautiful and/or evil female character rather than poisonings with the plant itself. A similar image is used in T. S. Elliot's poem *The Waste Land* (1913), where Belladonna is a seductress but, as we saw earlier, also a fitting plant to be found on wasteland.[48] There is even a DC Comics character called *Nightshade*, a female government super-spy and romantic interest for Captain Atom. A superstition associated with the plant is that at certain times it takes the form of a fatal enchantress of exceeding loveliness, whom it is dangerous to look upon.[20,49]

## 4.4   DEADLY NIGHTSHADE AND THE DEVIL

The witches of Greek mythology Hecate and Circe knew well the extremely poisonous and narcotic effects of this plant, and the first part of its name *atropa* is taken from the name of one of the fatal goddesses of the underworld, Atropos who cut the thread of life of mortals at a time of her choosing.[31,36,50,51] One of its folk names is "sorcerer's berry".[52] The second part *belladonna* comes from the Italian for "beautiful lady herb" as it was used by Renaissance Venetian ladies to enlarge the pupils of the eye, making them appear more seductive.[2,12] It is another candidate for the poison given to Odysseus' sailors (see Chapter 2). The medieval name for the plant, *dwale*, was thought to mean sleeping, trance, raging or deadly.[16,41]

Belladonna was one of the ingredients in something called "green unguent", an apparently foul-smelling concoction that witches used to help them fly.[53] In the 16th Century recipes for green unguent were published, and they included mandrake (Chapter 2), hemlock (Chapter 3), deadly nightshade and jimsonweed (Chapter 7), perhaps it's not surprising therefore that these plants are known as the "hexing herbs".[3,16,31] Belladonna also appears as a standard potion ingredient in *Harry Potter and the*

*Goblet of Fire* (2000).[54] Witches were also said to consume small amounts of deadly nightshade to aid in fortune telling.[11]

Another superstition was that it is one of the Devil's favourite plants, and he trims and tends to it by sprinkling it with his own blood.[11,20] On one night of the year, known as Walpurgis and celebrated on 30th April, it was safe to pick as the Devil was busy preparing for the festivities.[55] If the Devil was hanging around, releasing a black hen first would scare him away.[56] Although other storytellers said he left a 'nightmare monster' behind to guard it.[46] In another, children were told (presumably by anxious parents) that picking the berries would conjure up the Devil. Others thought that a sprig of deadly nightshade could be used to ward off evil spirits.[11]

### 4.5 MODERN MEDICAL USES

Atropine has a long history of legitimate medical use, but it is Datura plants (Chapter 7), that are grown to make medicinal atropine rather than belladonna.[19] As we have heard already, its ability to dilate our pupils has made it useful in specialist eye medicine; the berries themselves are no longer squeezed into the eye, as atropine was isolated from belladonna in 1819,[41] allowing us to create eye drops. Atropine is also used before surgery to reduce salivation,[57] and during emergency resuscitation for bradycardia (slow heart function associated with heart attack), as it causes our heart rate to increase.[31]

The whole belladonna plant is used in Traditional Chinese Medicine (TCM) to treat gastric ulcers, night sweats, drooling and asthma.[58] Only small amounts (up to 150 mg) of *Atropa belladonna* herb and root, can be sold in the UK by registered pharmacies and by, or under the supervision of, a pharmacist.[59] There are exemptions for herbalists to prescribe plant-based remedies within certain limits, although there is no mandatory training or registration for individuals wishing to set up as herbal practitioners.[28] This means that even when bought legally, use of deadly nightshade in herbal medicines can still be dangerous. In 2015 in the UK, a 50-year-old woman was found giggling, intoxicated and confused after drinking a herbal belladonna tincture for her insomnia. She was taken to hospital and fortunately recovered.[28]

Some homeopathic preparations made from belladonna and are recommended for paralysis,[60] meningitis,[22] inflammation including during breastfeeding,[49] and even for children who have had too much sun.[61] But the most high-profile use of deadly nightshade has been in homeopathic teething tablets in the USA. Teething preparations are used Worldwide by parents to soothe the pain suffered by babies when a new tooth erupts through their gums. Over a 10-year period (2006 to 2016) the Food and Drug Administration (FDA) received 370 reports of adverse effects from parents of babies given *Hyland's* homeopathic teething tablets or gels. Babies' eyes dilated, they had repeated seizures, became delirious, started twitching, stopped breathing and, in eight cases, died.[62] After an official warning from the FDA, the company stopped making the products, but like so many things that crop up in forensic toxicology cases, they remained on shop shelves and in bathroom cabinets for months afterwards. Unlike medicinal drugs, homeopathic products can be sold without any proof needed that they actually work or are safe. The idea that the ingredients are 'natural' can also lure parents into a false sense of security. We might not expect something homeopathic to be dangerous – given how heavily diluted the ingredients should be – but an inspection of the Hyland factory found poor quality control, with differing amounts of atropa belladonna in its products. Laboratory testing also showed that the amount on the label did not always match the contents.[62]

## 4.6 MODERN TOXICOLOGY CASES

Atropine is usually seen in toxicology laboratories in cases where the deceased was admitted to hospital or treated by paramedics just before their death. This means it mostly turns up in fatal motor vehicle crashes or heart attack victims.

Poisonings from eating the plant are relatively rare, but deadly nightshade berries are attractive and intensely sweet tasting,[55] and children are often tempted by them,[50] or they are mistaken for the fruits of other edible plants *e.g.*, bilberries (*Vaccinium myrtillus*).[17] Estimates of the number of belladonna berries needed to cause death vary from 1 to 30,[12,14] probably because the potency of the berries increases as the plant matures. A hospital in Turkey admitted 49 children with deadly nightshade poisoning between 1996 and 2003. Forty-eight of them had mistakenly eaten the plant, but one had been fed it by their parents as a remedy for diarrhoea (there have been several reports of parents giving children belladonna to cure various ailments[63]). The most common symptoms doctors saw were meaningless speech, lethargy and coma, and all of the children recovered.[64]

Another attraction for children is automatic atropine injector pens (like EpiPens®). These are very effective during emergencies as an antidote to chemical warfare agents, but do tend to go off accidentally in little hands, fortunately usually without fatal consequences.[65]

As we saw earlier, high temperatures do not affect atropine, so cooking doesn't destroy the toxins.[17] A couple in their 50s were taken to hospital in the UK in 1999 after being found at home in a confused state (they were grasping at imaginary objects and mumbling incoherently). The police noted that although the gas fire was burning in the house, the family dog was healthy, ruling out carbon monoxide poisoning from the fire (see Chapter 12).[17] Both of them had vomited after eating a homemade fruit pie, and the berries were identified by experts from Kew Gardens as being belladonna. Atropine was found in biological samples from both patients (and in the pie) and they remained in hospital for several days.[17]

Because of the overheating effect of belladonna, treatment requires wet cloths to reduce body temperature, and oxygen for breathing distress.[22]

As we have seen previously, the combination of herbal and prescription medications can make matters worse. In the case of belladonna, taking it alongside **quinidine**, a heart medication originally from the bark of the cinchona tree, can increase some of belladonna's effects, such as hallucinations.[22]

Fortunately there are antidotes, **pilocarpine** and **physostigmine**,[66] which can correct the lack of acetylcholine, and after administration and recovery most patients have no memory of the delirium.[11] It's important to realise that antidotes are not universally available, particularly in developing countries, so not everyone survives deadly nightshade poisoning, even with hospital care. Atropine itself is an antidote for poisoning by the alkaloid muscarine, which is found in certain mushrooms and the nerve agent sarin.[67]

Normal therapeutic concentrations of atropine depend on what it has been used for, but they range from 0.002–0.0025 mg L$^{-1}$ in serum.[65,68] In an unusual homicide from France in 2009, the body of a 51-year-old man was found decomposing in the boot of his car. He had been sawn in half in such a way that suggested his body was frozen first.[69] The atropine concentration in his peripheral post-mortem blood was 0.49 mg L$^{-1}$, consistent with a massive overdose of atropine. During the police investigation it was revealed that his wife had obtained three bottles of 1% atropine eye drops without a prescription in the month before his death.[70] The concentration of atropine in the man's cardiac blood was 0.89 mg L$^{-1}$, almost twice as high as the peripheral blood.[70] It is very common for forensic toxicologists to see higher concentrations of drugs and toxins in blood taken from the heart, than blood taken from what we call "peripheral" areas of the body. The most useful blood sampling site for toxicologists is the femoral vein, which can be found in your inside thigh. The higher heart concentrations are due to redistribution of chemicals in the body after death, and in the case of orally ingested poisons due to diffusion across the stomach lining into blood near the heart. Because of this, heart or "cardiac" blood drug concentrations are difficult to interpret, and this type of sample is only used when no other blood is available at the post mortem.[71] This is most likely to be the case if the person is heavily decomposed or lost a lot of blood prior to death, or in infants.

Like some of the other plants we have seen, there is still modern recreational abuse of atropine-containing plants, with some accidental poisonings in those attempting to get high.[36] In Switzerland between 1995 and 2009 there were 69 cases of this recorded by their Poisons Centre.[72] Atropine has also been found as a cutting agent in cocaine, and in 1997 a 39-year-old man was hospitalised in the USA with atropine poisoning after snorting adulterated cocaine.[57] Commercial asthma medications used to contain belladonna and these were widely abused, by ingesting the powder that was meant to be inhaled, until they were removed from the market.[63] People with asthma and bronchitis have also been known to smoke belladonna leaves and even ready-made belladonna cigarettes.[25,32] This might seem like an odd choice of remedy, but it causes the muscles of the respiratory system to relax, meaning that the tightness and swelling of the airway improves.[36]

Unlike hemlock (see Chapter 3) livestock and other mammals seem to develop a **tolerance** (see Chapter 2) to deadly nightshade, and so poisonings in animals are rare.[14] Although, rabbits are not affected by atropine, as they have the enzyme atropinesterase, which deactivates it,[17] humans have been poisoned after eating rabbit meat and even honey tainted with atropine.[11,73]

## 4.7  CATCHING THE CHEMICAL CULPRIT

Deadly nightshade plant extracts and biological fluids from hospital patients can be analysed using **GC-MS** (see Chapter 3), however the very high temperatures involved can convert atropine to scopolamine, making it harder to pin down which chemical (and plant) was responsible for the symptoms. **LC-MS** (see Chapter 2) is a better choice, as this analysis is done at room temperature.[74]

Before any kind of instrument can be used on a biological sample in forensic toxicology, a clean-up process called an "extraction" is needed. This pulls any drugs or poisons out of the samples into a clean solvent such as ethyl acetate. Apart from keeping expensive instruments clean, it also makes it easier to narrow down signals of interest. If we injected a blood or urine sample straight into the machine, it would detect all sorts of naturally present chemicals in the body that have not caused the death. Of course, some naturally occurring compounds, such as ethanol, cyanide (Chapter 10) and carbon monoxide (Chapter 12) can also be deadly, so we have to be careful not to exclude them. Extraction is often the most time-consuming part of a forensic toxicology case, with many tests taking a whole day to complete. Others have multiple steps or need to be left overnight to dry. Labs try to be as efficient as they can by "batching" cases together, *i.e.*, waiting until several cases that need to be tested for atropine have been submitted, before starting the lab work, but this can lead to delays.

A technique used in deadly nightshade toxicology cases is **liquid–liquid extraction** (**LLE**).[75] It uses the fact that organic solvents (which are "hydrophobic") and water (or mainly watery biological samples, which are "hydrophilic") don't mix well, and also the fact that most drugs or poisons would rather be in the solvent than the blood or urine. To tempt them into the organic liquid, a change in the pH of the sample is required – that means it needs to be made more acidic or more alkaline (depending on the drugs we are looking for). Once the acid or base has been added, the solvent (*e.g.*, hexane or diethyl ether) sits in a layer on top of the sample and the two are mixed together as much as possible, using shaking or stirring. Although the two layers never actually disappear, the line between them (the "interface") can become blurry. To clear it up, the samples are spun in circles at very high speed in a machine called a "centrifuge". Once the interface is clear, the organic solvent is removed using a pipette, and the watery layer is thrown away. It might seem odd to throw away a forensic sample, but it is the drugs we are interested in, not the biological material itself (*e.g.*, red blood cells). The organic solvent is then injected into the GC-MS or LC-MS instrument. If there's a large volume of liquid (several mL) it is usually dried down (or "concentrated") first under nitrogen gas (to avoid any reaction with oxygen) and then "reconstituted" (the residue re-dissolved) in another solvent such as ethyl acetate (a few µL). This helps us to detect low concentrations of drugs. For some drugs, another step is needed at this point, known as **derivatisation** (see Chapter 10).

## 4.8 CASE CLOSED

In our case, the stage directions are for Juliet to fall gracefully and immediately asleep on her bed.[1] As we have seen, those consuming deadly nightshade would typically go through some of the other symptoms first *e.g.*, racing heart, flushed face, confusion and agitation before falling into a coma.[32] Like anything that is taken orally, symptoms would also take time to come on, hours in some of the real-life cases we have seen, although taking a potent tincture would undoubtedly be quicker than eating the plant. It is also difficult to find modern evidence of a belladonna coma lasting as long as 42 h, as the Friar says (typically it would be around 10 h[76]). However as Juliet drinks the potion shortly before 3 am on Wednesday and awakes just before dawn on Thursday, her coma actually only lasts about 27 h,[77] more in keeping with the 30 h coma we saw earlier in one of the real-life cases.

Today we understand the action of deadly nightshade in molecular detail, and we can now easily compare Juliet's symptoms to real poisoning cases. But writers such as Shakespeare relied on herbals like Gerard's and Banckes's for their poisonous plots. Although Juliet could have taken any of the poisons in Chapters 2 and 3, which were common knowledge at the time, the effects are closest to Gerard's description "dead sleep" of deadly nightshade.

## REFERENCES

1. W. Shakespeare, in *Comedies*, David Campbell Publishers Ltd, London, 2000, vol. 1, ch. 5, pp. 408–511.
2. M. Willes, *A Shakespearean Botanical*, Bodleian Library, Oxford, 2020.
3. J. Mann, *Murder Magic and Medicine*, Oxford University Press, Oxford, 1992.
4. A. S. Harper-Leatherman and J. R. Miecznikowski, *J. Chem. Educ.*, 2012, **89**, 629–635.
5. E. Tabor, *Econ. Bot.*, 1970, **24**, 81–94.
6. K. Harkup, *Death by Shakespeare: Snakebites, Stabbings and Broken Hearts*, Bloomsbury, London, 2020.
7. *Gerard's Herbal*, ed. M. Woodward, Senate, London, 1994.
8. A. C. Kail, *Med. J. Aust.*, 1983, **2**, 515–519.
9. T. Hargreaves, *Poisons and Poisonings: Death by Stealth*, RSC Publishing, Cambridge, 2017.
10. A. Börsch-Haubold, *Sci. Sch.*, 2007, **4**, 50–55.
11. M. R. Lee, *J. R. Coll. Physicians Edinburgh*, 2007, **37**, 77–84.
12. G. Grigson, *The Englishman's Flora*, Helicon, Herzelia, 1996.
13. D. Frohne and H. J. Pfänder, *Poisonous Plants: A Handbook for Doctors, Pharmacists, Toxicologists, Biologists and Veterinarians*, Manson Publishing, London, 2nd edn, 2004.
14. P. M. North, *Poisonous Plants and Fungi*, Blandford Press, London, 1967.
15. M. R. Cooper and A. W. Johnson, *Poisonous Plants and Fungi in Britain*, The Stationery Office, London, 1998.
16. S. Schultes, A. Hofmann and C. Rätsch, *Plants of the Gods*, Healing Arts Press, Rochester, VT, 2001.
17. H. J. Southgate, M. Egerton and E. A. Dauncey, *J. R. Soc. Promot. Health*, 2000, **120**, 127–130.
18. J. H. Bock and D. O. Norris, in *Forensic Plant Science*, ed. J. H. Bock and D. O. Norris, Academic Press, San Diego, 2016, ch. 1, pp. 1–22.
19. S. Funayama and G. A. Cordell, *Alkaloids: A Treasury of Poisons and Medicines*, Elsevier, Amsterdam, 2015.
20. Anonymous, *The Poison Garden*, The Alnwick Garden, Alnwick, 2005.
21. A. Chevallier, *Encyclopedia of Medicinal Plants*, DK Publishing, St Leonards, 2001.

22. Medical Economics Company, *PDR for Herbal Medicines*, Medical Economics Company, Montvale, NJ, 1998.
23. Prescription Cost Analysis - England, https://www.nhsbsa.nhs.uk/statistical-collections/prescription-cost-analysis/prescription-cost-analysis-england-2019, 2020, accessed June 2021.
24. Royal Pharmaceutical Society, *British National Formulary*, BNF Publications, London, 2021.
25. W. Sneader, *Drug Discovery: A History*, John Wiley & Sons, Chichester, 2005.
26. H. C. Williams and A. Du Vivier, *Contact Dermatitis*, 1990, **23**, 119–120.
27. K. A. Marquardt and R. S. Tharratt, *Clin. Toxicol.*, 1994, **32**, 75–78.
28. A. Chadwick, A. Ash, J. Day and M. Borthwick, *BMJ Case Rep.*, 2015, **2015**, bcr2015209333.
29. M. Chamberlin, *Old Wives' Tales: The History of Remedies, Charms and Spells*, The History Press, Cheltenham, 2020.
30. *Principles of Forensic Toxicology*, ed. B. Levine, AAAC Press, Washington DC, 2013.
31. J. M. Riddle, *Goddesses, Elixirs, and Witches*, Palgrave MacMillan, New York, 2010.
32. A. Demirhan, Ü. Y. Tekelioğlu, İ. Yıldız, T. Korkmaz, M. Bilgi, A. Akkaya and H. Koçoğlu, *Turk. J. Anaesthesiol. Reanim.*, 2013, **41**, 226–228.
33. F. Schneider, P. Lutun, P. Kintz, D. Astruc, F. Flesch and J.-D. Tempe, *J. Toxicol., Clin. Toxicol.*, 1996, **34**, 113–117.
34. *A New Grove Dictionary of Opera*, ed. S. Sadie, Oxford University Press, Oxford, 1997.
35. G. Weitz, *Br. Med. J.*, 2003, **327**, 1469–1471.
36. K. Fatur, *Econ. Bot.*, 2020, **20**, 1–19.
37. Tristan und Isolde Libretto, https://www.opera-arias.com/wagner/tristan-und-isolde/libretto/, accessed July 2020.
38. J. L. Müller, *J. Toxicol., Clin. Toxicol.*, 1998, **36**, 617–627.
39. J. Keats, *John Keats The Major Works*, Oxford University Press, Oxford, 2001.
40. C. W. King, in *The Encyclopedia of Ancient History*, ed. R. S. Bagnall, K. Brodersen, C. B. Champion, A. Erskine and S. R. Huebner, Blackwell, Oxford, 2013, ch. 16, pp. 5586–5587.
41. E. A. Campbell, *Victorian Lit. Cult.*, 2007, **35**, 607–615.

42. C. S. Hornfeldt and J. E. Collins, *J. Toxicol., Clin. Toxicol.*, 1990, **28**, 185–192.
43. M. O'Leary, in *The Anthology of English Folk Tales*, ed. N. Guy, The History Press, Stroud, 2016, ch. 7, pp. 45–52.
44. W. Wordsworth, *The Poems of William Wordsworth: Collected Reading Texts from The Cornell Wordsworth*, Humanities Press, Penrith, 2009.
45. G. Dunne, *Practical Magic Screenplay*, Warner Brothers, Burbank, CA, 1998.
46. F. Inkwright, *Botanical Curses and Poisons*, Liminal 11, London, 2021.
47. J. J. Roche, *Life of John Boyle O'Reilly: Together with His Complete Poems and Speeches*, T. Fisher Unwin, London, 1891.
48. P. Sicker, *Twent. Century Lit.*, 1984, **30**, 420–431.
49. M. Castro, *The Complete Homeopathy Handbook: A Guide to Everyday Health Care*, Macmillan, London, 1990.
50. D. Frohne and H. J. Pfänder, *A Colour Atlas of Poisonous Plants*, Wolfe Publishing Ltd, London, 1983.
51. A. Stewart, *Wicked Plants*, Algonquin Books of Chapel Hill, Chapel Hill, NC, 2009.
52. S. T. Dietz, *The Complete Language of Flowers: A Definitive and Illustrated History*, Wellfleet Press, New York, 2020.
53. R. Richardson, *Britain's Wild Flowers: A Treasury of Traditions, Superstitions, Remedies and Literature*, The National Trust, London, 2017.
54. J. K. Rowling, *Harry Potter and the Goblet of Fire*, Bloomsbury, London, 2000.
55. M. Grieve, *A Modern Herbal*, Dover Publications, New York, 2nd edn, 1971.
56. F. Inkwright, *Folk Magic and Healing: An Unusual History of Everyday Plants*, Liminal 11, London, 2019.
57. A. L. Weiner, M. J. Bayer, C. A. McKay Jr, M. DeMeo and E. Starr, *Am. J. Emerg. Med.*, 1998, **16**, 517–520.
58. J. Zhou, G. Xie and X. Yan, *Encyclopedia of Traditional Chinese Medicines*, Springer, Heidelberg, 2011.
59. Banned and restricted herbal ingredients, https://www.gov.uk/government/publications/list-of-banned-or-restricted-herbal-ingredients-for-medicinal-use/banned-and-restricted-herbal-ingredients, accessed May 2020.

60. R. Root-Bernstein and M. Root-Bernstein, *Honey, Mud, Maggots and Other Medical Marvels*, Houghton Mifflin Company, New York, 2000.
61. Homoeopathic Development Foundation, *Homoeopathy for the Family*, Wigmore Publications Limited, London, 1988.
62. Hundreds of Babies Harmed by Homeopathic Remedies, Families Say, https://www.scientificamerican.com/article/hundreds-of-babies-harmed-by-homeopathic-remedies-families-say/, accessed July 2021.
63. K. Fatur and S. Kreft, *Toxicon*, 2020, **177**, 52–88.
64. H. Çaksen, D. Odabaş, S. Akbayram, Y. Cesur, Ş. Arslan, A. Üner and A. F. Öner, *Hum. Exp. Toxicol.*, 2003, **22**, 665–668.
65. *Clarke's Analysis of Drugs and Poisons*, ed. A. C. Moffatt, D. Osselton and B. Widdop, Pharmaceutical Press, London, 4th edn, 2011.
66. L. S. Nelson, M. A. Howland, N. A. Lewin, S. W. Smith, L. R. Goldfrank and R. S. Hoffman, *Goldfrank's Toxicologic Emergencies*, McGraw Hill, New York, 11th edn, 2018.
67. D. Johnson, in *Chemistry in its Element*, RSC Publishing, Cambridge, 2013.
68. M. Schulz, A. Schmoldt, H. Andresen-Streichert and S. Iwersen-Bergmann, *Crit. Care*, 2020, **24**, 195.
69. M. Brown, *Death in the Garden: Poisonous Plants & their Use Throughout History*, Pen & Sword Books Ltd, Barnsley, 2018.
70. J. Carlier, E. Escard, M. Péoc'h, B. Boyer, L. Romeuf, T. Faict, J. Guitton and Y. Gaillard, *J. Forensic Sci.*, 2014, **59**, 859–864.
71. R. W. Byard and D. M. Butzbach, *Forensic Sci. Med. Pathol.*, 2012, **8**, 205–207.
72. J. Fuchs, C. Rauber-Lüthy, H. Kupferschmidt, J. Kupper, G.-A. Kullak-Ublick and A. Ceschi, *Clin. Toxicol.*, 2011, **49**, 671–680.
73. R. Bevan-Jones, *Poisonous Plants: A Cultural and Social History*, Windgather Press, Oxford, 2009.
74. J. Beyer, O. H. Drummer and H. H. Maurer, *Forensic Sci. Int.*, 2009, **185**, 1–9.
75. L. Dufayet, E. Alcaraz, J. Dorol, C. Rey-Salmon and J.-C. Alvarez, *Forensic Toxicol.*, 2020, **38**, 264–268.
76. B. Hubbard, *Poison: The History of Potions, Powders and Murderous Practitioners*, Welbeck, London, 2020.
77. W. Shakespeare, *Romeo and Juliet*, Cambridge University Press, Cambridge, 2003.

# CHAPTER 5

# The Queen of Poisons

If you see a term that's **bold** it's defined in the Glossary. Only the first time that the word appears in the chapter will it be indicated in this way.

---

**Case History: Apparent Suicide**

*After hearing of the death of his teenage girlfriend, a young man approaches an apothecary asking for a fatal poison in order to take his own life. A few hours later he is found dead alongside her.*

ROMEO:      ...A dram of poison, such soon-speeding gear
            As will disperse itself through all the veins
            That the lide-weary taker may fall dead,
            And that the trunk may be discharged of breath
            As violently as hasty powder fired
            Doth hurry from the fatal cannon's womb.

Act V, scene i

*The Tragedy of Romeo & Juliet* by William Shakespeare,[1] 1594

---

Poisonous Tales: A Forensic Examination of Poisons in Fiction
By Hilary Hamnett
© Hilary Hamnett 2023
Published by the Royal Society of Chemistry, www.rsc.org

## 5.1 THE INVESTIGATION

Although this case is from the same play as our case study in Chapter 4, Romeo's self-poisoning was intended to be fatal rather than the elaborate plot device of Juliet's. In modern terminology we would describe this as a "completed" suicide.

Romeo uses a different source of poison to Juliet, who gets hers from the friar, this time it's an apothecary – a person who prepared and sold drugs and remedies.[2] In Shakespeare's time there were no pharmacies or dispensing chemists, but the apothecary in *Romeo & Juliet* (1594) is supposed to sell medicines, not poisons (the latter having been outlawed in the region due to a spate of revenge poisonings). As we have seen in previous chapters, the line between medicine and poison can be very fine indeed and in fact has a special name – the "therapeutic index" (or "margin of safety"). If a substance has a wide therapeutic index such as paracetamol, it is considered safer than one with a narrow therapeutic index such as digoxin (see Chapter 6). The apothecary is rather cagey about the ingredients, telling Romeo only

APOTHECARY:     Put this in any liquid things you will
                And drink it off, and if you had the strength
                Of twenty men, it would dispatch you straight.

Act V, scene i
*The Tragedy of Romeo & Juliet* by William Shakespeare,[1] 1594

In Shakespeare's time, there was such a poor understanding of how dose is related to toxicity, that an apothecary could sell the same underlying substance as a cosmetic, cure or poison.[2]

Two authors have recently suggested that Romeo's poison contained either cyanide (see Chapter 10) or an extract of the plant aconite, rather than the deadly nightshade Juliet may have used (see Chapter 4).[2-5] Both were known in Shakespeare's time thanks to the various published herbals. Gerard tells us that all of the many (over 200) different varieties of aconite are deadly and venomous,[6] and Banckes and Turner say that a concoction of aconites can be deadly to beasts such as wolves.[7,8] We also meet this theme in the form of the "wolfsbane potion" in *Harry Potter and the Prisoner of Azkaban* (1999),[9] and *Harry Potter and the Half-Blood Prince* (2010).[10] Although deadly to wolves, there are some species of caterpillars that can feed happily on aconite plants without ill effects.[11] The plants are definitely deadly to humans however, and were used for centuries as arrow or sword poisons or to poison enemy water supplies in times of war.[12–15]

## 5.2 THE PLANT BEHIND THE STORY

Aconite (*Aconitum napellus*) is a member of the buttercup family and goes by many different names including "monkshood", "Devil's helmet", "Queen of poisons" and "wolfsbane".[16] It also has the dubious honour of being the most poisonous plant in Europe. The plant has bright green 'toothy' leaves (Figure 5.1, right), and violet–blue flowers with a helmet- or cowl-shaped hood (Figure 5.1, left) similar to that of a monk's habit.[5,17,18] The flowers are seen in the UK in May to June,[19] and are particularly attractive to bees (see Figure 5.1).[20] The seeds are tiny, black and triangular, and the plant likes to grow in wet and shady places such as on the banks of streams, particularly in alpine regions,[17] but also in Southwest England.[19,21]

**Figure 5.1**   (Left) The flowers of an aconite plant being visited by a bee, taken at the Chelsea Physic Garden. (Right) The leaves of the aconite plant, taken at The Poison Garden in Alnwick.

All parts of the plant are exceedingly poisonous as they contain the alkaloids **aconitine, mesaconitine, hypaconitine** and **jesaconitine**.[17,22] The main chemical culprit is aconitine,[23] known as a "secondary compound" because it is not needed to keep the plant alive, but is used as a defence mechanism – which is why secondary compounds often taste unpleasant.[24] The usual source of poison is the dried root,[2] which is harvested in the autumn,[25] and remains toxic as the alkaloids survive the drying process.[19] Unfortunately the root can often be mistaken for wild radishes.[2] Soaking or boiling in water leads to hydrolysis of the aconite alkaloids, which can make them less toxic.[26] The root is at its most toxic during winter.[17] Dioscorides tells us that the root looks similar to the tail of a scorpion and if given to a scorpion, makes it insensible until it is given another plant, hellebore (*Helleborus niger*) (Figure 5.2) as a cure.[27] Far from being a cure, hellebore is itself poisonous and one of its alkaloids, **protoanemonin**, causes tinging in the throat and mouth, vomiting and cardiac symptoms.[28]

**Figure 5.2**    A black hellebore plant in flower, taken at The Poison Garden in Alnwick.

Aconitine is also found in the plant's fresh leaves,[17] which are at their most potent in early summer just before flowering, and have been mistaken for a kind of edible grass[29,30] (known as *Momijigasa* in Japan[31]) or parsley.[19] As with many poisonous plants, the potency depends on the species, place of origin and the time it is picked,[29] but only a few grams of the plant are dangerous due to how little of the alkaloids are required to produce toxic symptoms.[17,32]

Aconite acts differently on the body to the Solanaceae plants from the previous chapters. It interferes with the way sodium moves between cells in the body,[33] increasing the flow rate[34] and allowing non-sodium metal ions (such as potassium and calcium) to sneak through undetected.[18] These sodium ions are crucial to keeping the muscles, heart and CNS running, and we will see how these are affected by aconitine shortly.[35]

In classical times, aconite was used as a pain-relieving solution for the eyes or was added to lumps of meat to kill beasts such as panthers.[27] A recipe where it was mixed with toasted cheese was recorded for killing mice and rats.[36] It was widely used in an epidemic of homicides in Ancient Rome, and an aconite draught was used to euthanise old and infirm men in Ancient Greece.[18] Because its narrow therapeutic index made it difficult to get the dosage right, it has fallen out of medical use apart from in Traditional Chinese Medicine (TCM) (see Section 5.5).

During the 19th Century, aconite became a valued pain reliever for neuralgia and sciatica in Europe and the United States.[37,38] Tinctures of aconite (made with alcohol, as aconite doesn't dissolve well in water[39]) were applied with a brush and absorb well through the skin,[17,22] in fact even just touching the plant can lead to numbness in the fingertips.[40] As well as helping to dissolve the alkaloids, the alcohol also increases the absorption of aconitine by enhancing vasodilation (opening up of blood vessels).[41] These tinctures were apparently still in use for pain in Ireland in the 1880s, as an overdose of aconite is mentioned in James Joyce's novel *Ulysses* (1922). The novel is about Leopold Bloom, whose father Rudolph took his own life by self-poisoning with a liniment (lotion or balm) containing aconite.

The Queen's Hotel, Ennis, county Clare where Rudolph Bloom (Rudolf Virag) died on the evening of the 27 June 1886, at some hour unstated, in consequence of an overdose of monkshood (aconite) self-administered in the form of a neuralgic liniment composed of 2 parts of aconite liniment to 1 of chloroform liniment.

Chapter 17
*Ulysses* by James Joyce,[42] 1922

It's worth noting that **chloroform**, which was also in the liniment, is very toxic too, so the death in this case is likely due to the effects of both compounds. Best known as an anaesthetic, it also had an early history as a poison and a way of incapacitating victims.[43] This complication is something forensic toxicologists are faced with most of the time; people rarely overdose on just one drug or poison.

When aconite is taken by mouth, the first signs of poisoning are a tingling or burning sensation in the mouth, then in the fingers and toes, as more and more of the body's nerve endings are affected by the aconitine.[22,33] This is followed by the feeling of ants crawling under the skin,[44] bouts of sweating and feeling icy cold, which progress to diarrhoea, accompanied by agonising vomiting, intense pain and paralysis. With fatal doses, breathing becomes irregular and difficult[19] and the heartbeat slows down. This interference with the normal pattern of the heart's beating (known as "tachycardia")[17,45] can be mistaken by doctors for digitalis (Chapter 6) poisoning if the victim goes to hospital. Fortunately poisoning by aconitine can be treated successfully with drugs such as flecainide or amiodarone[46] which act as metal ion channel blockers (to slow down the flow of ions between cells), as well as with β-blockers.[45]

Treatment is not always successful however, and death can follow within 6 h because the muscles of the lungs or heart become paralysed.[17] Taken orally, the effects of aconitine come on within 10–20 min.[17,22] Unlike the plants we met in previous chapters, which brought on hallucinations and confusion, aconite generally leaves the mind clear.[19]

### 5.3 POISONOUS PLOTS

Poisoning with aconite may go back as far as the Apocrypha in the Bible, with the story of the High Priest Alcimus (or Alkimos) who died supposedly of a stroke after making an unpopular decision about the Temple in Jerusalem. He wanted to remove a wall to give non-priests access to the inner sanctuary

Now in the one hundred and fifty-third year, in the
second month, Alcimus gave orders to tear down the wall of the inner
court of the sanctuary.
He tore down the work of the prophets!
But he only began to tear it down, for at that time Alcimus was
stricken and his work was hindered;
his mouth was stopped and he was paralyzed, so that he could no
longer say a word or give commands concerning his house.
And Alcimus died at that time in great agony.

1 Maccabees 9: 54–56
*The New Revised Standard* Bible[47]

Although some of the symptoms described are typical of a stroke (loss of speech, paralysis, collapse and sudden death[48]), the severe pain or "great agony" is an unusual one. As we have seen, intense pain *is* associated with aconite poisoning however.[44] We also know that monkshood plants were common the Mediterranean region, and this combined with the political turmoil at the time has led modern sources to suggest Alcimus was actually a victim of deliberate poisoning.[48]

The poisoning of the monarch in Shakespeare's *King John* (1596) on the battlefield by the monk he had employed as his taster, is described as starting with a fever, but quickly escalating

KING JOHN:     Poison'd, ill fare; dead, forsook, cast off:
               And none of you will bid the winter come
               To thrust his icy fingers in my maw,
               Nor let my kingdom's rivers take their course
               Through my burn'd bosom, nor entreat the north
               To make his bleak winds kiss my parched lips
               And comfort me with cold...

Act V, scene vii
*King John* by William Shakespeare,[49] 1596

Although Shakespeare uses uncharacteristic detail to describe the poisoning, neither the identity of the poison nor the motive is ever revealed. However, some of the effects (sweating and a burning sensation) are consistent with the poison being aconitine, but that would not explain the hallucinations he later suffers. As in real cases dealt with by toxicologists, it's possible more than one poison was used.[2]

Aconite is mentioned by name in another of Shakespeare's plays, *Henry IV, Part 2* (1597). The king, who is unwell, is fighting off a rebellion whilst his eldest son Hal prepares to succeed him. Keen to ensure a peaceful succession to the throne, he urges one of his other sons to support Hal,[5] giving advice on keeping the band of brothers together. Aconite is used as a byword for poison or venom and compared to gunpowder. The brothers should resist the rumours and criticism that will follow after the King's death

HENRY:      ...Mingled with venom of suggestion—
            As, force perforce, the age will pour it in—
            Shall never leak, though it do work as strong
            As aconitum or rash gunpowder.

Act IV, scene iv
*History of Henry IV, Part 2* by William Shakespeare,[49] 1597

It is this analogy with gunpowder that we also saw in *Romeo & Juliet* (where the poison is called the "fatal cannon") that has convinced some interpreters that Shakespeare meant Romeo's poison to be aconite.[5]

In the poetic novel *Our Lady of the Flowers* (1943) by Jean Genet, the main character Divine recalls that as a boy he used to gather aconite leaves from the garden, roll them up and swallow them, each time increasing the dose.[50] This behaviour is reminiscent of King Mithridates VI of Pontus (in modern-day Turkey) who lived from 135–63 BCE, and was so paranoid about attempts to remove him from the throne that he starting consuming various poisons to make himself immune. This gave birth to a legend about a special mixture, called a "mithridate", which could act as an antidote to any poison. We will meet Mithridates again in Chapter 8 on **strychnine**.

In Roman times, the extremely bitter wormwood (Figure 5.3) was thought to be an antidote to aconite poisoning,[51] but we now know there is no specific antidote for this plant.[52,53]

In Ben Jonson's play *Sejanus His Fall* (1603) about Aelius Sejanus, a powerful advisor behind the Roman emperor Tiberius Caesar, we see aconite being used as a metaphor for setting one schemer on another. Tiberius leaves Rome, and asks Marco (the aconite) to spy on Sejanus (the scorpion). Both are plotting to take the throne

**Figure 5.3**    Roman wormwood, taken at The Tuner Garden.

TIBERIUS:          ...I have heard that aconite,
                   Being timely taken, hath a healing might
                   Against the scorpion's stroke; the proof we'll give:
                   That while two poisons wrestle, we may live.

                                                    Act III, scene iii
                               *Sejanus His Fall* by Ben Jonson,[54] 1603

We also meet some old sleep-inducing poisons in the previous scene of the play. Mandrake and poppy (Chapter 2) and hemlock (Chapter 3)

SEJANUS:           ...And may they lay that hold upon thy senses,
                   As thou hadst snuft up hemlock, or ta'en down
                   The juice of poppy and of mandrakes. Sleep,
                   Voluptuous Caesar,...

                                                    Act III, scene ii
                               *Sejanus His Fall* by Ben Jonson,[33] 1603

Oscar Wilde's *Lord Arthur Savile's Crime* (1891) features an attempted aconite poisoning. Lord Arthur is looking for the perfect poison to get rid of his wife, Lady Clemetina, so he can marry another woman

> ...in the second volume of Erskine, he found a very interesting and complete account of the properties of aconitine, written in fairly clear English. It seemed to him exactly the poison he wanted. It was swift–indeed, almost immediate, in its effect – perfectly painless,...

> p. 23
> *Lord Arthur Savile's Crime* by Oscar Wilde,[55] 1891

He persuades the pharmacist that he needs the poison for a rabid dog and tells his wife that it is a homeopathic remedy for heartburn. The account of aconite's effects Lord Savile reads is not very accurate; as we have seen, intense pain is one of the symptoms of aconitine poisoning, so it is lucky that Clementina never takes the capsule of poison, and dies of natural causes.

More recently, wolfsbane flowers featured in mass poisonings in the Playstation game *Ghost of Tsushima* (2020) about a samurai warrior in 13th Century Japan.

Cleopatra may have killed herself with a poisonous cocktail of aconite, hemlock (Chapter 3) and opium poppy (Chapter 2) rather than with the bite of a cobra. We will explore this intriguing theory in Chapter 9.

### 5.4   WOLFSBANE AND WITCHCRAFT

Aconite has been known since classical times, and Hecate the Greek goddess of witchcraft and mistress of the art of poisoning, knew of its deadly properties.[18,33] Its first mention in classical Greek myth tells us that it came from the saliva of the three-headed dog Cerberus,[33] and the story can be found in Ovid's *Metamorphoses*. This time the potion mixer is the beautiful witch Medea (Figure 5.4), wife of Aegeus.

> Now Theseus, whose father Aegeus had never known him, arrived;
> his heroic deeds had established peace on the Isthmus of Corinth.
> Bent on his murder, Medea prepared him a potion of aconite,
> brought by her earlier over the sea from the Scythian shores.
> The poison is said to have come from the teeth of the monstrous
> dog whom Echídna bore.

> Book 7, 404–409
> *Metamorphoses* by Publius Ovidius Naso,[56] 8th Century AD

**Figure 5.4**    Medea, by Frederick Sandys.

Medea is trying to kill Theseus because she is worried she will lose her influence over Aegeus if his son appears on the scene. Fortunately, Theseus is saved by his father who recognises his sword, and hence his son, and stops him drinking the poison.[57] The more familiar character of Hercules also features in this story, as he was the one to capture Cerberus and bring him up from the underworld.[18,58] Hercules was said to have fought the dog on the hill of Aconitus.[59]

Aconite may also be the poison used in an earlier book of *Metamorphoses* where the goddess Athena sprinkles her rival Arachne with a magical juice, turning her into a spider. Translations vary, but some believe the "baleful herb" to be aconite[60]

...With that she departed, sprinkling the girl with the magical juice
of a baleful herb. As soon as the poison had touched Arachne,
Her hair fell away, and so did the ears and the nose.
The head now changed to a tiny ball and her whole frame shrunk in
proportion.

Book 6, 138–142
*Metamorphoses* by Publius Ovidius Naso,[56] 8th Century AD

Aconite has kept its magic and mystery through the centuries: In medieval times, witches were rumoured to use aconite in their preparations for orgies and flying about,[37] and it can be found in the first potions lesson of *Harry Potter and the Philosopher's Stone* (1997),[61] where it is mentioned by Professor Snape. In India, aconite is sacred to blue-throated Shiva, Hinduism's God of All Poisons.[62] According to legend, Shiva drank a poison that was threatening the World, but was saved from death by his wife who choked his throat causing it to turn blue. A few drops of the poison that missed his mouth fell to the ground as aconite.[53] In Nepal a mandrake-like legend appeared that digging up aconite roots releases evil spirits imprisoned within the plants, inflicting dire calamity on the digger.[63]

Hellebore is also associated with folklore and magic; white hellebore was believed to protect against sorcery, whereas black hellebore brought bad luck on anyone who picked it.[64] Hellebores were also said to be protected by eagles.[53]

### 5.5 MODERN MEDICAL USES

Aconitum species have been important in TCM for over 2000 years, with the tubers and roots being the most common ingredients.[65,66] They are used to treat various illnesses such as pain, inflammation and heart problems,[22] but also as a kind of 'diet therapy'.[34] Aconite is a key ingredient in a homemade medicinal liquor or aconitum wine – monkshood root in white spirit – and three deaths from drinking this were recorded in China in 2017.[65] In another case in 2021, it had actually been prescribed by a doctor to treat shoulder pain. Aconite mixed with alcohol makes it more soluble, and so easier to absorb.[67] In six more deaths associated with these liquors in China in 2011, one man drank his tonic for 18 days in a row, and another drank 8 times the recommended amount.[34] Taking too much medication is a surprisingly common story for forensic toxicologists – unfortunately we can sometimes expect miracles from our medicines, and people become impatient if their symptoms do not clear up immediately and so take another few doses. The same thing can happen to recreational drug users. Accidental overdoses at music festivals happen when someone takes a party pill like ecstasy and expects the effects to come on straight away. They don't because it takes time for the pill to dissolve in the stomach and pass into the bloodstream. In the meantime the festival-goer takes another pill and when the effects do start, the combination can lead to a dangerous speeding up of the body's processes, causing them to overheat.

Medicines containing aconite roots are also available as capsules and tablets,[68] and in France in 2005, a 21-year-old man was admitted to A&E, though survived, after taking three homemade aconite capsules for phytotherapy (using plant-derived medications to treat and prevent disease). He made the capsules by grinding up dried tubers, and was taking them to calm his anxiety. Although he had survived taking one capsule a day for several months, he decided to take three at once to increase the effects. He suffered heart problems, chest pain and vertigo.[69]

When used correctly in TCM, the aconite root is first "processed" (boiled) to reduce the toxic alkaloid content by "hydrolysing" the active chemicals (reacting them with water to make less toxic versions).[70,71] A 35-year-old man in Hong Kong died after drinking aconite broth in 1994, having been told by the prescriber to boil it for only 40 min, when the recommended time is actually 2 h.[71] Some unqualified doctors in rural areas may prescribe raw aconite, whose side-effects are made worse by the lack of local medical facilities if a patient does suffer poisoning.[34]

In Ayurveda (Hindu traditional medicine) aconite root (known as "Vatsanabha") is used to treat ailments such as asthma, arthritis and skin disorders.[72] It should only be used after the aconite is processed in a way known as "Shodhana", the instructions for which seem to vary widely, meaning it is not always successful. The process generally involves soaking the root in cow's urine for three days, followed by boiling it in cow's milk and exposure to the sun.[73] If the purification is done incorrectly, or the patient does not follow medical advice, they can overdose on these aconite preparations, as happened to a 47-year-old man in India in 2009, who took four Ayurveda tablets containing aconite for abdominal pain and ended up in hospital.[73]

Countermeasures for aconite poisoning include gastrointestinal emptying, magnesium and calcium infusions, **atropine** and **lidocaine** for getting the heart back on track and artificial respiration (using a ventilator).[22]

Aconite can be freely purchased from herbal shops in many places around the World,[12] and used without a prescription.[23]

In the UK, the import, manufacture, supply or sale of unlicensed herbal medicines containing aconite are banned.[74] Aconite can be legally used in homeopathic remedies (as these should not contain any active ingredient).[15,22,75]

## 5.6   MODERN TOXICOLOGY CASES

The most common causes of poisoning with aconite plant material are either a suicide attempt or incorrect preparation of the plant for TCM,[12] with most fatalities being reported in East Asia for this reason.[29,76] However there have been poisonings in North America and Europe. In 2002 in the UK, a 60-year-old male drank an aconite extract he made by liquidising the plant, in a suicide attempt. He was taken to A&E but died following a heart attack.[32,35] Two middle-aged men died in Germany in 2020 after also drinking aconite extracts.[41]

In the USA in 2005, a 59-year-old woman was taken to A&E after accidentally taking an unknown root called "Hmong medicine #9". She had a racing irregular pulse and low blood pressure, but recovered after 8 h.[45] In Germany in 2019, an elderly man died after drinking an unlabelled home-made aconite plant extract that was inside a vodka bottle.[35] Drinking mis-labelled liquids is another story familiar to most forensic toxicologists; people will sometimes re-use old soft drink or even medicine bottles to store poisons such as pesticides and leave the original label on the outside.

In Canada in 2008, a 25-year-old ate some wild flowers during an afternoon walk. Within 4 h he had collapsed and was unable to be resuscitated.[46] Although this collapse could be due to any number of medical reasons, it is often seen in patients who have consumed a poison that interferes with the heart.

As well as the deaths in China we saw in Section 5.4, poisonings also happen in other countries in East Asia. An 81-year-old woman died in A&E in Korea in 2019 the day after drinking a liquid that had been heat extracted from the root of the aconite plant. Within an hour of ingesting the liquid she began to feel nauseous and experienced stomach pain and eventually died of cardiac arrest.[23]

In Singapore in 2015, a 34-year-old man ended up in A&E after preparing and drinking his own TCM broth. He had incorrectly copied down the weights of ingredients from a recipe on the internet, adding 10 times the safe dose of aconite root.[77] Fortunately, he recovered after a week in hospital, but it is not just members of the public who can make dosage errors; forensic toxicologists will sometimes have to deal with cases where a doctor or nurse made the error in hospital.

It is also not only those attempting to self-medicate who can run into problems, there have also been a number of fatalities caused by people eating aconite root in order to get 'high' (although aconite is not an hallucinogen).[17]

If faced with a potential aconite poisoning, it is not always clear to the forensic toxicologist, which of the deadly alkaloids to look for in the body. Although **aconitine** seems to be responsible for the main effects of the plant, it is not very stable once in the body and so only very small amounts may be detectable after consumption.[32] We can look for other alkaloids instead; for example, in Japan in 2007 there were five cases of aconite poisoning and **jesaconitine** was the main chemical found in four of them, with **mesaconitine** in the fifth. **Hypaconitine** was detected only in one case.[78] In other reports from toxicologists, **mesaconitine** and **hypaconitine** are either not detected at all or seen only in trace amounts.[23] This is probably because the victims ate different parts of the plant such as the roots or leaves, or may have eaten plants grown in different areas or climates.[78]

There is also the question of *where* to look in the body. Case reports have shown that the greatest concentration of aconite alkaloids occurs in the urine,[32,79] and that they can be seen in urine long after they have disappeared from the blood.[78] Finding a drug or poison in the urine tells us that at some point in the past, the person took it, but it cannot tell us *when*. And some drugs (such as cannabis) can lurk in the urine for weeks. The amount of the poison in the urine can be measured, but the number doesn't really mean anything. It cannot be related to dose or symptoms, or used to declare the cause of death.

A much more useful sample in poisoning cases where the victim dies in hospital is an ante-mortem blood or serum. These are usually taken on admission to A&E when doctors are still investigating the cause of the patient's symptoms. Although hospitals can do some testing for drugs and poisons, they are often slow and limited, and in any case, emergency doctors' focus is usually on treating life-threatening symptoms rather than finding out exactly what has caused them. If the ante-mortem samples are kept by the hospital (not always the case, as hospitals have limited storage space) they can be sent to the forensic toxicologist and analysed alongside the post-mortem samples. In our case from Korea earlier, the concentration of **aconitine** in the ante-mortem serum was 0.0391 mg $L^{-1}$, but this had fallen to 0.0286 mg $L^{-1}$ in the post-mortem peripheral blood.[23] This is likely due to the patient metabolising the aconitine whilst they were still alive. Serum concentrations can be misleading however, because of the way the sample is prepared. A normal whole blood sample (Figure 5.5, right) is drawn and then placed into a special tube containing a gel. The gel is inert so it does not react chemically with the blood, but acts as a separating layer when the sample is centrifuged. After spinning, the heavier red blood

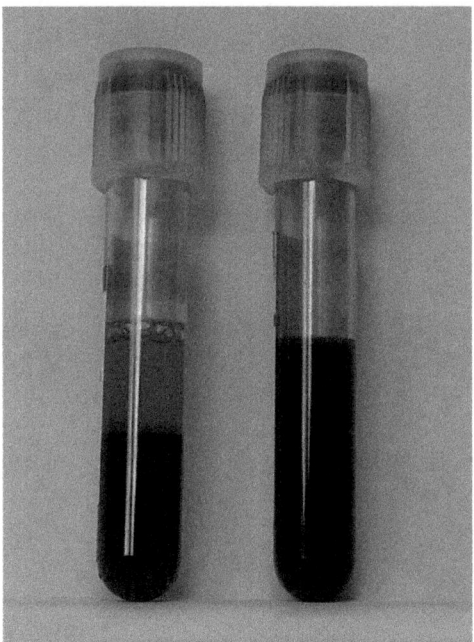

**Figure 5.5** Serum (left) and whole blood (right). Reproduced from https://commons.wikimedia.org/wiki/File:Blut-EDTA.jpg, under the terms of the CC BY-SA 3.0 license, https://creativecommons.org/licenses/by-sa/3.0/deed.en.

cells collect at the bottom of the tube underneath the gel, and the lighter (straw coloured) serum sits on top (Figure 5.5, left).

Unfortunately, drugs and poisons like to stick to the gel layer and so serum samples can give falsely low readings. In another twist, the gels contain toluene, which can also be found in the blood of people who have been sniffing solvents such as glue or petrol, so this can be another misleading result. Checking in samples of the brain and lungs (assuming these were taken at the autopsy) for toluene can help to distinguish the glue sniffers.

In a weird homicide case, a man in his 50s was found dead behind the steering wheel of his car beside a highway in Belgium.[26] The autopsy revealed signs of strangulation, but the post-mortem toxicology testing using **high-performance liquid chromatography with diode array detection (HPLC–DAD)** (Chapter 7) was negative. The case remained unsolved for five years until his widow confessed to mixing a decoction of aconitum leaves and stalks into a bottle of red wine. She claimed to then use a rope to move the 'body', leading to the signs of strangulation, although the Pathologist took the view that the man was still alive but in a coma at this point. The toxicology lab had fortunately kept the biological samples, which is often the case in unsolved homicides, and when re-tested using the more sensitive **LC-MS** (Chapter 2) they found aconitine.[26,80] The red wine likely made the decoction even stronger, as we know that alcohol enhances the toxicity of aconitine by making the cell walls more permeable to metal ions.[34]

Although re-testing was successful in this case, it is a rare example; biological samples often degrade over time *e.g.*, blood will clot leading to uneven distribution of drugs or poisons throughout the liquid. To further complicate matters, even if the samples survive intact, the drugs or poisons themselves may degrade, as they are broken down into other chemicals or metabolites by microbes in the blood. A lot depends on the storage temperature *e.g.*, in a fridge at 5 °C or a freezer at −20 °C and if any preservatives were added to the sample containers. The most common preservative is sodium fluoride (NaF). This helps to stop microbes growing and keep drug levels stable. It's not just the breakdown of drugs that causes problems years later however, some chemicals, such as alcohol[81] and the date-rape drug GHB,[82] can actually be produced in post-mortem samples over time. Even with preservatives and freezing, you are not guaranteed that the results from later testing will truly represent those at the time of death. The number of times a sample is taken out of the freezer, thawed and then re-frozen can also affect drug stability. In the case of toxic gases such as carbon monoxide (Chapter 12), just taking off the lid of the sample affects the levels in the blood. In criminal cases there will be a written or electronic record of who has handled the samples at each stage, from autopsy to receipt at the laboratory, to the final toxicology test. This is known as the "chain of custody".

## 5.7 CATCHING THE CHEMICAL CULPRIT

As we saw in Chapter 4, drugs and poisons need to be extracted from biological samples by forensic toxicology labs. Aconitine can be extracted from blood using a separating technique known as **solid-phase extraction (SPE)**.[12] This makes use of a kind of mini filter known as a "cartridge", which lets through all of the biological material, such as red blood cells, whilst trapping the drugs or poisons. A solvent such as methanol is then passed through the filter and this picks up the drugs and carries them out of the cartridge. A vacuum is used to help pull the liquids through, as gravity on its own is too slow. This method gives us a small volume of solvent containing just the chemicals of interest, which can then be analysed using an instrument. As we saw earlier, only trace amounts of the poisonous alkaloids may be present in the blood after death, therefore a very sensitive method such as **LC-MS** is needed to detect them.[26]

## 5.8 CASE CLOSED

Deaths from aconite poisoning are somewhat difficult to diagnose, because, like many poisons, there are often no obvious signs on the body of trauma or disease.[34] Even if poisoning is suspected, there are also no tell-tale clues after death that aconite specifically was the culprit.[23] Sometimes, signs of asphyxia can be seen at the post-mortem,[83] but this could be due to several causes of death, not least being suffocated.

Shakespeare's death for Romeo is as rapid as Juliet's, with him falling into death (or at least unconsciousness) immediately after drinking the poison.[1] Although death from aconite usually takes hours to occur, there was a case in China in 2011 where the victim died within 10 min, so Romeo's death is not that unrealistic.[34] At the time, Turner described it as the "most hasty poison".[84] It's worth remembering too that many of the slower fatal cases we have seen in this chapter involved some kind of medical treatment that prolonged the life of the patient. Taking aconite as a liquid also means it is absorbed more rapidly than eating the plant itself.[23]

When Juliet awakes from her feigned death, she tries to poison herself for real this time, but seeing Romeo has drunk all of the liquid, tries catch the last few drops by kissing his lips

JULIET:     Poison, I see, hath been his timeless end.
            O chur! Drunk all, and left no friendly drop
            To help me after? I will kiss thy lips.
            Haply some poison yet doth hang on them
            To make me die with a restorative.

Act V, scene iii
*The Tragedy of Romeo & Juliet* by William Shakespeare,[1] 1594

It's unlikely that enough aconite would have remained on Romeo's lips for a fatal dose, but this was a common plot device used in other plays and novels of the time.

## REFERENCES

1. W. Shakespeare, in *Comedies*, David Campbell Publishers Ltd, London, 2000, ch. 5, vol. 1, pp. 408–511.
2. K. Harkup, *Death by Shakespeare: Snakebites, Stabbings and Broken Hearts*, Bloomsbury, London, 2020.
3. A. S. Harper-Leatherman and J. R. Miecznikowski, *J. Chem. Educ.*, 2012, **89**, 629–635.
4. D. I. Macht, *Bull. Hist. Med.*, 1949, **23**, 186–194.
5. M. Willes, *A Shakespearean Botanical*, Bodleian Library, Oxford, 2020.
6. *Gerard's Herbal*, ed. M. Woodward, Senate, London, 1994.
7. Anonymous, *Herbal*, Rycharde Banckes, London, 1525.
8. W. Turner, *A New Herball*, Cambridge University Press, Cambridge, 1989.
9. J. K. Rowling, *Harry Potter and the Prisoner of Azkaban*, Bloomsbury, London, 1999.

10. J. K. Rowling, *Harry Potter and the Half-Blood Prince*, Bloomsbury, London, 2010.
11. J. H. Bock and D. O. Norris, in *Forensic Plant Science*, ed. J. H. Bock and D. O. Norris, Academic Press, San Diego, 2016, ch. 1, pp. 1–22.
12. J. Beyer, O. H. Drummer and H. H. Maurer, *Forensic Sci. Int.*, 2009, **185**, 1–9.
13. Anonymous, *The Poison Garden*, The Alnwick Garden, Alnwick, 2005.
14. E. Tabor, *Econ. Bot.*, 1970, **24**, 81–94.
15. M. Castro, *The Complete Homeopathy Handbook: A Guide to Everyday Health Care*, Macmillan, London, 1990.
16. T. Hargreaves, *Poisons and Poisonings: Death by Stealth*, RSC Publishing, Cambridge, 2017.
17. D. Frohne and H. J. Pfänder, *A Colour Atlas of Poisonous Plants*, Wolfe Publishing Ltd, London, 1983.
18. J. Mann, *Murder Magic and Medicine*, Oxford University Press, Oxford, 1992.
19. P. M. North, *Poisonous Plants and Fungi*, Blandford Press, London, 1967.
20. M. Grieve, *A Modern Herbal*, Dover Publications, New York, 2nd edn, 1971.
21. G. Grigson, *The Englishman's Flora*, Helicon, Herzelia, 1996.
22. Medical Economics Company, *PDR for Herbal Medicines*, Medical Economics Company, Montvale, NJ, 1998.
23. Y. S. Cho, H.-W. Choi, B. J. Chun, J. M. Moon and J.-Y. Na, *Forensic Sci., Med., Pathol.*, 2020, **16**, 330–334.
24. S. R. Whitehead and M. D. Bowers, *Am. Nat.*, 2013, **182**, 563–577.
25. A. Chevallier, *Encyclopedia of Medicinal Plants*, DK Publishing, St Leonards, 2001.
26. A. A. Van Landeghem, E. A. De Letter, W. E. Lambert, C. H. Van Peteghem and M. H. A. Piette, *Int. J. Legal Med.*, 2007, **121**, 214–219.
27. P. Dioscorides, *The Greek Herbal of Dioscorides*, Haner Publishing, New York, 1959.
28. D. Frohne and H. J. Pfänder, *Poisonous Plants: A Handbook for Doctors, Pharmacists, Toxicologists, Biologists and Veterinarians*, Manson Publishing, London, 2nd edn, 2004.
29. Y. Gaillard and G. Pepin, *J. Chromatogr. B: Biomed. Sci. Appl.*, 1999, **733**, 181–229.

30. *Clarke's Analysis of Drugs and Poisons*, ed. A. C. Moffatt, D. Osselton and B. Widdop, Pharmaceutical Press, London, 4th edn, 2011.
31. N. Yoshioka, K. Gonmori, A. Tagashira, O. Boonhooi, M. Hayashi, Y. Saito and M. Mizugaki, *Forensic Sci. Int.*, 1996, **81**, 117–123.
32. S. P. Elliott, *Sci. Justice*, 2002, **42**, 111–115.
33. R. Highfield, *The Science of Harry Potter: How Magic Really Works*, Headline Book Publishing, London, 2002.
34. Q. Liu, L. Zhuo, L. Liu, S. Zhu, A. Sunnassee, M. Liang, L. Zhou and Y. Liu, *Forensic Sci. Int.*, 2011, **212**, e5–e9.
35. V. Hofmann, A. Landmann, G. Schmitt, A. Krauskopf and M. Bartel, *Forensic Toxicol.*, 2020, **38**, 511–516.
36. M. Brown, *Death in the Garden: Poisonous Plants & their Use Throughout History*, Pen & Sword Books Ltd, Barnsley, 2018.
37. A. Been, *Pharm. Hist.*, 1992, **34**, 35–39.
38. A. C. Kail, *Med. J. Aust.*, 1983, **2**, 515–519.
39. B. Clegg, in *Chemistry in Its Element*, RSC Publishing, Cambridge, 2016.
40. M. R. Cooper, A. W. Johnson and E. A. Dauncey, *Poisonous Plants and Fungi: An Illustrated Guide*, The Stationery Office, Norwich, 2nd edn, 2003.
41. F. Veit, M. Gürler, A. Nebel, C. G. Birngruber, R. B. Dettmeyer and W. Martz, *Forensic Sci. Int. Rep.*, 2020, **2**, 100158.
42. J. Joyce, *Ulysses*, Penguin, London, 1986.
43. J. F. O'Brien, *The Scientific Sherlock Holmes: Cracking the Case with Science & Forensics*, Oxford University Press, New York, 2013.
44. F. Inkwright, *Botanical Curses and Poisons*, Liminal 11, London, 2021.
45. S. W. Smith, R. R. Shah, J. L. Hunt and C. A. Herzog, *Ann. Emerg. Med.*, 2005, **45**, 100–101.
46. R. Pullela, L. Young, B. Gallagher, S. P. Avis and E. W. Randell, *J. Forensic Sci.*, 2008, **53**, 491–494.
47. *The New Oxford Annotated Bible (New Revised Standard Version)*, ed. M. D. Coogan, M. Z. Brettler, C. A. Newsom and P. Perkins, Oxford University Press, Oxford, 2006.
48. F. P. Moog and A. Karenberg, *Adverse Drug React. Toxicol. Rev.*, 2002, **21**, 151–156.

49. W. Shakespeare, *The History of King Henry the Fourth Part 2*, Clarendon, Oxford, 1998.
50. J. Genet, *Our Lady of the Flowers*, Grove Press, New York, 1991.
51. R. Bevan-Jones, *Poisonous Plants: A Cultural and Social History*, Windgather Press, Oxford, 2009.
52. L. S. Nelson, M. A. Howland, N. A. Lewin, S. W. Smith, L. R. Goldfrank and R. S. Hoffman, *Goldfrank's Toxicologic Emergencies*, McGraw Hill, New York, 11th edn, 2018.
53. S. Lawrence, *Witch's Garden: Plants in Folklore, Magic and Traditional Medicine*, Welbeck, London, 2020.
54. B. Jonson, in *The Devil is An Ass and Other Plays*, ed. M. J. Kidnie, Oxford University Press, Oxford, 2000, ch. 3, pp. 103–221.
55. O. Wilde, *Lord Arthur Savile's Crime and Other Stories*, La Spiga, Milan, 1996.
56. M. Ovid, *Metamorphoses*, Penguin, London, 2004.
57. C. M. Skinner, *Myths and Legends of Flowers, Trees, Fruits, and Plants: In All Ages and in All Climes*, J. B. Lippincott, Philadephia, 1911.
58. A. Stewart, *Wicked Plants*, Algonquin Books of Chapel Hill, Chapel Hill, NC, 2009.
59. B. Hubbard, *Poison: The History of Potions, Powders and Murderous Practitioners*, Welbeck, London, 2020.
60. E. W. Hesse, *Hispania*, 1952, **35**, 74–82.
61. J. K. Rowling, *Harry Potter and the Philosopher's Stone*, Bloomsbury, London, 1997.
62. G. S. Bause, *Anesthesiology*, 2015, **123**, 376.
63. C. J. S. Thompson, *Poison Romance and Poison Mysteries*, George Routledge & Sons, Ltd, London, 1904.
64. F. Inkwright, *Folk Magic and Healing: An Unusual History of Everyday Plants*, Liminal 11, London, 2019.
65. M. Pan, X. Wang, Y. Zhao, W. Liu and P. Xiang, *Forensic Sci. Int.*, 2019, **298**, 39–47.
66. Y. Liu, in *History of Toxicology and Environmental Health*, ed. P. Wexler, Academic Press, Boston, 2015, ch. 9, pp. 89–97.
67. Y. Ya, Z. Zhixiang, L. Chao, Z. Wei, W. Zhiyong, C. Huafeng, Z. Shaohua and X. Hongfei, *J. Forensic Sci.*, 2021, **66**, 2035–2040.
68. T. Y. K. Chan, *Clin. Toxicol.*, 2009, **47**, 279–285.
69. F. Moritz, P. Compagnon, I. G. Kaliszczak, Y. Kaliszczak, V. Caliskan and C. Girault, *Clin. Toxicol.*, 2005, **43**, 873–876.

70. D. Csupor, E. M. Wenzig, I. Zupkó, K. Wölkart, J. Hohmann and R. Bauer, *J. Chromatogr. A*, 2009, **1216**, 2079–2086.
71. P. Dickens, Y. T. Tai, P. P. H. But, B. Tomlinson, H. K. Ng and K. W. Yan, *Forensic Sci. Int.*, 1994, **67**, 55–58.
72. P. K. Sarkar, P. K. Prajapati, V. J. Shukla and B. Ravishankar, *Toxicol. Int.*, 2012, **19**, 35–41.
73. A. K. Panda and S. K. Debnath, *Int. J. Ayurveda Res.*, 2010, **1**, 183–186.
74. Banned and restricted herbal ingredients, https://www.gov.uk/government/publications/list-of-banned-or-restricted-herbal-ingredients-for-medicinal-use/banned-and-restricted-herbal-ingredients, accessed May 2020.
75. Homoeopathic Development Foundation, *Homoeopathy for the Family*, Wigmore Publications Limited, London, 1988.
76. Y. G. Zhang and G. Z. Huang, *Am. J. Forensic Med. Pathol.*, 1988, **9**, 313–319.
77. S. Sheth, E. C. C. Tan, H. H. Tan and L. Tay, *Singapore Med. J.*, 2015, **56**, e116–e119.
78. Y. Fujita, K. Terui, M. Fujita, A. Kakizaki, N. Sato, K. Oikawa, H. Aoki, K. Takahashi and S. Endo, *J. Anal. Toxicol.*, 2007, **31**, 132–137.
79. C.-K. Lai, W.-T. Poon and Y.-W. Chan, *J. Anal. Toxicol.*, 2006, **30**, 426–433.
80. J. Beike, L. Frommherz, M. Wood, B. Brinkmann and H. Köhler, *Int. J. Legal Med.*, 2004, **118**, 289–293.
81. F. C. Kugelberg and A. W. Jones, *Forensic Sci. Int.*, 2007, **165**, 10–29.
82. F. P. Busardò and A. W. Jones, *Clin. Toxicol.*, 2019, **57**, 149–163.
83. S. Shreya and D. Ganapathy, *Drug Invent. Today*, 2018, **11**, 677–681.
84. W. Turner, *A New Herball*, Cambridge University Press, Cambridge, 1551.

CHAPTER 6

# Outfoxed by Digitalis

If you see a term that's **bold** it's defined in the Glossary. Only the first time that the word appears in the chapter will it be indicated in this way.

---

**Case History: Political Poisoning**

*British Secret Service agent James Bond is sent to Madagascar to pursue a man called Le Chiffre who is known to finance terrorist organisations. Learning that Le Chiffre plans to raise money in a high-stakes poker game in a casino, Bond sets out to win, but during the game takes a sip of a poisoned martini. His vision begins to blur, he becomes sweaty, dizzy, unsteady and vomits. Panicking, he staggers to his Aston Martin and pulls out a surgical needle, stabs himself in the wrist and attaches the needle to an electronic box. This instantly calls the MI6 medics' room who can see that he is going into cardiac arrest.*

DOCTOR #2:      Do we know what it is yet?
DOCTOR #1:      Still scanning

*A green screen running through a database suddenly comes up with a match*

DOCTOR #2:      Ventricular tachycardia. Digitalis.
                     What the hell do we give him? The kit has amphetamines,
                     antihistamines—
                     Lidocaine! That'll work

*(continued)*

---

Poisonous Tales: A Forensic Examination of Poisons in Fiction
By Hilary Hamnett
© Hilary Hamnett 2023
Published by the Royal Society of Chemistry, www.rsc.org

> *Bond stabs himself with a medi pen containing lidocaine, then passes out before activating the defibrillator. [His love interest] Vesper arrives in time to fire the defibrillator and Bond jolts and immediately regains consciousness.*
>
> Lines 149–162
>
> *Casino Royale* a screenplay by Neal Purvis and Robert Wade,[1] 2005

## 6.1 THE INVESTIGATION

Slipping something into your enemy's drink seems like the oldest trick in the book – and we might expect James Bond to be a tad more careful. In the film we see a vial of liquid poison being poured into Bond's glass, which would take longer than the few seconds on screen to start having an effect. However, the symptoms Bond displays paint an accurate picture of digitalis poisoning.

Spiked drinks are most often seen by forensic toxicologists in alleged drug-facilitated sexual assault cases. The same techniques we use to analyse blood and urine samples can be used on drinks,[2] but they are rarely available as evidence. This is because it usually takes some time for the complainant to notice the effects, by which time the drink has either been finished or cleared away. Many spiked drinks go unnoticed unless they have a strong aftertaste (as is experienced with the date-rape drug GHB, which tastes salty). In the film *Casino Royale*, Bond looks at his martini, realises poison is the likely culprit but leaves the glass at the table. Preserving evidence may be the priority for a forensic scientist, but Bond's most immediate problem is dealing with being poisoned.

The MI6 doctors realise Bond is suffering from ventricular tachycardia. This is a kind of very fast heartbeat, which means that there's not enough time to fill the heart with blood between beats. Although the script of the movie tells us he is in cardiac arrest, this is a bit premature – cardiac arrest means that the heart has stopped altogether. Bond's attempt to treat his ventricular tachycardia with a defibrillator could have worked, but could also have made things much worse; shocking someone not in cardiac arrest can further complicate the electrical signals in the heart.[3]

Despite the injection of **lidocaine** (see Section 6.3) Bond's heart condition does progress to cardiac arrest, and now a defibrillator *is* needed. This is because the electrical activity in his heart has become chaotic, stopping it from pumping. Instead, it quivers or "fibrillates" and needs a controlled electric shock from a defibrillator to restart its normal beat.[4] All of this drama has not dealt with the underlying cause of his symptoms however, being poisoned. In cases of digitalis poisoning, the patient needs to be given the DigiFab® antidote (see Section 6.2) to prevent their heart problems from coming back.[5] Not for our hero though. Bond's recovery from his cardiac arrest is so speedy that he's able to return to the poker game, whereas most people would be taken to intensive care or even put into a coma to allow their bodies time to recover.[4]

The most interesting part of this case study to a forensic toxicologist however is MI6's machine, which can identify the poison in a person's blood who is thousands of miles away within 40 s. In reality, even at top speed it would take a few days to clean up the blood sample and analyse it for something like digitalis. There are some more rapid tests available, like the ones used in hospitals for diagnosis of poisoning, but even these take hours and cannot give us the name of the specific poison. At best they can tell us which group or 'family' of drugs has been used.

### 6.2  THE PLANT BEHIND THE STORY

Foxgloves (*Digitalis*) are a group of around 20 plants, the most familiar of which is *Digitalis purpurea* or common foxglove. It is also known as "dead men's bells", "ladies' glove",[6] "thimble flower" or "finger flower".

The plants often stand tall with lance-shaped leaves arranged in a rosette[7] and bell-shaped purple, white or pink flowers (Figure 6.1),[8,9] with the pink ones being the most common in the UK.[10] Usually the lowest flowers on the stalk open first.[11] Foxgloves like acidic soils and shade, and are often found in open spaces in woodland.[10,12] The flowers are distinctive, but the plant does not flower in its first year, and the poisonous leaves can be mistaken for edible plants such as borage, comfrey[13] or kale.[14] Comfrey tea is used to treat various aches and pains, but even if you do manage to pick the correct plant, comfrey (*Symphytum*

**Figure 6.1**   (Left) A pink foxglove plant, taken in Northwood. (Right) A white foxglove plant, taken at The Turner Garden showing the lower flowers open first.

*officinale*) contains **retronecine**, which has been linked to liver cancer, so is probably best avoided anyway.[15]

Digitalis plants flower in June and July,[16] and are a common sight in the UK. The plant tastes hot-bitter and has a slightly unpleasant odour,[6] alerting many a would-be victim to their poisoner's intent. Despite the taste, accidental poisonings have been reported when people have tried to suck the nectar out of the flowers.[17] Foxglove contains what we call **cardiac glycosides**, including **digoxin**, **digitoxin**, and **digitoxigenin** (with digoxin being the most potent).[8,18] You may have heard of these as prescription heart medications, because they can rapidly strengthen the heartbeat.[8] Cardiac glycosides are found in some other plant species, including oleander (Chapter 7) and Lily of the Valley (*Convallaria majalis*).[19]

Although foxglove plants contain these alkaloids, getting digoxin and digitoxin out of them is a complicated process.[6] The leaves contain 0.1–0.4% cardiac glycosides,[5] and are most potent before flowering.[20] Bales of dried foxglove leaves are shipped to chemical processing plants and chopped up before being extracted with a solvent. The whole process is not very efficient; it takes about 1000 kg of dried foxglove leaves to make 1 kg of pure digoxin.[21] The plants themselves also give very variable results, as like any herbal medication, the potency depends on the climate, season and age.

For those using herbal remedies, once ground down, foxglove leaves look pretty generic; in 2010 in China there was an outbreak of poisoning caused by some tea that a group of nine neighbours thought was made of comfrey (*S. officinale*) a well-known herbal medicine,[13] but it was in fact digitalis. Without the distinctly different flowers, the plants look similar, and the foxglove leaves had been picked and brewed by accident. Unfortunately, boiling the leaves does not reduce their toxicity,[22] and all nine were admitted to hospital, but recovered.

We saw in Chapter 5, that some plants and medicines have a very narrow therapeutic index, and digitalis is another example. Because the plants themselves vary so widely in cardiac glycoside content, it is safer to use known amounts of the pure alkaloids.[6] Unlike many medicines, which are made entirely synthetically from pure chemical starting materials ("precursors"), prescribed digoxin still comes from the foxglove plant.[21]

Digoxin acts in two ways on the heart – first it increases the strength of heart muscle action (by increasing the amount of blood pumped on each beat[23]) and also decreases the power of the electrical signals that control the heart's activity.[10] This latter effect might not seem like a good idea, but it can help if the electrical pulses within the heart have become irregular. It is this effect that makes foxglove so deadly however, as too much of it will supress the electrical signals altogether, bringing the heart to a standstill. Similar to the **aconitine** we met in Chapter 5, digoxin interferes with the movement of sodium and calcium ions between nerve cells. Calcium ions can start to build up in the cells making the heart pump harder to push blood out. In appropriate doses, digoxin can correct pumping irregularities, which is why it is a valuable medicine.[3] Doses that are measured to be therapeutic can bring on unintended side-effects straight away such as loss of appetite, vomiting, abdominal pain, sweating, diarrhoea and headache.[6,14] When digitalis is taken, it also interferes with an enzyme in the eyes producing a yellowish sheen or 'halo' on the patient's vision, particularly when looking at a light (as is seen in *Casino Royale*).[10] With increasing doses, cardiac rhythm disorders, which can be life-threatening, develop, along with confusion and hallucinations.[6] Death is usually due to heart failure.

Digitalis has a long history of use as a folk medicine, thought to have started in Ireland. During the 1500s and 1600s it had some medicinal uses (such as being boiled in wine and drunk for treating 'naughty humours') but its effect on the heart was not understood.[16] Later, it was a common remedy sold by apothecaries (see Chapter 5),[24] for conditions like ulcers, boils, headaches, abscesses, paralysis and high blood pressure.[6] A tincture of digitalis (made with 25% alcohol)[6] was used as a cardiac tonic and stimulant.

In the 18th Century, a doctor called William Withering decided to take a closer look at digitalis.[8,25] He discovered that the plant could help those suffering from abnormal fluid build-up, or "dropsy" as it was called back then.[21,26] Withering spotted the connection between scarlet fever and strep throat (both diseases that can damage the heart), the fluid building up in the patient, and his foxglove cure.[21] We now know that fluid build-up can be a side-effect of heart problems.[17] Foxglove can have two important effects on patients with heart disease, where the heart can't maintain normal circulation. As we saw earlier, it can help the heart to beat more strongly, slowly and regularly without the need for more oxygen. At the same time it ramps up urine production, by increasing blood flow to the kidneys,[27] and this lowers the volume of fluid in the body and eases the load on the heart.[8]

### 6.3  POISONOUS PLOTS

Despite the striking and common form of the foxglove plant on the UK landscape, it was not mentioned by name by Shakespeare or any of the Old English poets.[28] It's possible that this was due to having a limited geographic distribution in the UK at this point in history.[29] However, in the novel *Silas Marner* by George Eliot (1861)[30] we see a chronically ill woman with heart disease, Sally Oates, who is swollen with fluid, being cured with an infusion of foxgloves by Silas himself. Even in the early 19th Century, locals see his ability to cure her as a sign of witchcraft or Devil worship,[31] and his reluctance to charm away the whooping cough of the local children leads to Silas becoming even more isolated in his small community than he was already. Understanding of herbal medicines was still poor during this period and it wasn't until the late 1800s that the chemical digoxin was isolated from the plant material, and it was the 1960s before the chemical structure was identified.[10]

MI6's treatment for Bond's ventricular tachycardia, or abnormal heartbeat, with lidocaine would work to slow down the heart's contractions, but would not tackle the underlying cause – the digoxin. This can be removed by first emptying the stomach (Bond tries this himself by drinking some salty water) and then injecting the patient with an antibody to digoxin called DigiFab® or digoxin Fab.[14] Antibodies are large Y-shaped proteins that seek out and match up with a specific target – in this case digoxin. When it recognises its target floating around in the blood, the antibody sticks itself (or "binds") to it, stopping it from affecting the body. Once bound to the antibody, the digoxin can no longer interfere with the heart, and symptoms usually start to subside after about 30 min. Interestingly, digoxin levels in the blood reduce more quickly than the symptoms, something that forensic toxicologists see with other drugs too; the level of drug in the blood doesn't always correlate with the effects the user experiences. Antibodies are the basis of a type of preliminary or "presumptive" test used in forensic toxicology (see Section 6.5).

The downside to DigiFab® is its cost (approx. £1000 per vial), so fortunately there is another antidote, **cholestyramine**, which speeds up the removal of digoxin by the body's natural elimination processes. As we saw above, the effect of too much digoxin is to slow down the electrical signals in the heart, so another plant poison **atropine** (Chapter 3) can be given to speed the heart up again. This clearly becomes a delicate balancing act for the doctors treating the patient.

We see foxglove being used for this purpose in the novel *Precious Bane* by Mary Webb (1924).[32] In the story, a family of farmers uses foxglove to lower the pulse of a cow whose heart seems to be bursting (we are told the cow is given a "dose" of foxglove leaves but are given no more detail). It also has a more sinister role in the tale, in the death of an elderly lady, who is given foxglove tea to drink after she becomes a burden. Interestingly, the "bane" in the title of the book doesn't refer to the poison, but to the cleft palate the main character, Prue Sarn, has. In the days of the Napoleonic wars, when the book is set, this was a sign of witchcraft. There were also stories of poor women drinking foxglove tea in the 1880s as a cheaper way to get intoxicated than alcohol.[33] Although feelings of drunkenness are not reported by dogixin users, it has been used as an hallucinogen.

Wordsworth immortalised the foxglove in his play *The Borderers* (1842) about clashes between those living on either side of the Scottish Border during the reign of Henry III (during the 1200s). In one of the scenes a homeless woman recounts a (pretty plausible) dream

BEGGAR:     I've had the saddest dream that ever troubled
            The heart of a living creature.—My poor Babe
            Was crying, as I thought, crying for bread
            When I had none to give him; whereupon,
            I put a slip of foxglove in his hand,
            Which pleased him so, that he was hushed at once:
            When, into one of those same spotted bells
            A bee came darting, which the Child with joy
            Imprisoned there, & held it to his ear,
            And suddenly grew black, as he would die.

Act I
*The Borderers: A Tragedy* by William Wordsworth,[34] 1842

Foxgloves are a favourite of honeybees,[28] and they are well known for crawling into European foxglove bells in order to pollinate them.[22] In areas where foxglove has been introduced (such as South America), the flowers have evolved to entice hummingbirds in as pollinators instead.[35] Tennyson features foxglove, alongside yew (Chapter 3) in his poem *In Memoriam* (1850)

Bring orchis, bring the foxglove spire,
The little speedwell's darling blue,

83
*In Memoriam A.H.H.* by Alfred Tennyson,[36] 1850

Although dismissed by some as "far-fetched",[31] there has been a suggestion that Ophelia in Shakespeare's *Hamlet* (1600) might have been depressed and died of eating foxglove leaves. The flowers aren't named, but long purple flowers were included in her garland, as described by Hamlet's mother, Queen Gertrude

GERTRUDE: There with fantastic garlands did she come
Of crow-flowers, nettles, daisies, and long purples
That liberal shepherds give a grosser name,
But our cold maids do dead men's fingers call them.

Act IV, scene vii
*The Tragedy of Hamlet, Prince of Denmark* by
William Shakespeare,[37] 1600

Ophelia is described as being found floating in a stream, and so the Queen assumes she has deliberately drowned herself. But there are other possibilities, such as that she lost her footing and fell in accidentally.[38] Today if someone is suspected of drowning, water would be found in their lungs at the autopsy and there are tests for certain chemical markers (*e.g.*, sodium and magnesium ions). These tell us whether the person was dead before they entered the water.[39] It's possible that Ophelia fell into the water after eating something poisonous, but foxgloves are not called "dead men's fingers". A couple of other plants go by this name: the first is *Decaisnea insignis* (Figure 6.2), which has finger-shaped edible fruit. The second is hemlock water dropwort or *Oenanthe crocata*, which contains the very poisonous alkaloid **oenanthotoxin**.[22,40] More information on hemlock plants can be found in Chapter 3. Another theory is that Shakespeare was referring to the early-purple orchid (*Orchis mascula*).[41]

### 6.4 FOXGLOVES AND FAIRIES

There is a strong association of foxgloves with magic, especially fairies. Names for the plant include fairy...fingers, gloves, caps, hats and thimbles.[6] Some legends say fairies live in the bells of the flowers.[33] Others say that should you wish fairies to visit your garden, do not pick the flowers as it causes them offence.[42] Fairies may be attracted to foxgloves and leave a mark of their visit as a white spot inside the 'bells' of the flowers. Garden centres in the UK still sell "Fairy seed" mixes, which sometimes contain foxglove seeds.[43] If you wanted to rid yourself of troublesome fairies, Turner suggests putting some foxglove stalks, leaves and flowers between your horse's back and saddle.[44] It was important to pick them under the full moon as that is when they are at their most

**Figure 6.2**    The Decaisnea insignis plant. Reproduced from https://commons.
                wikimedia.org/wiki/File:5206-Decaisnea_fargesii-20111103-ham-
                burg.jpg, under the terms of the CC BY-SA 3.0 license, https://
                creativecommons.org/licenses/by-sa/3.0/deed.en.

powerful.[45] Other legends said it was unlucky to bring foxgloves,
especially white ones, into the house.[46]

In 1851, a "fairy doctress" was convicted of giving a child foxglove. When the paralytic child was discovered, she claimed that
the real one had "been stolen by fairies".[31] This might sound
like a bizarre claim, but in folklore, newborn babies that failed
to thrive were sometimes suspected of being "changelings" –
sickly fairy babies left in place of healthy human ones. The test
for a real human baby was bathe it in the juice of a foxglove
plant (or give it a drink of foxglove juice) and see if it survived;
the fairy babies died, whilst the human babies would start to
thrive.[33,47]

The legend of foxglove may extend beyond fairies to Roman
mythology. In Ovid's version of the birth of Mars given in his book
*Fasti*, an unnamed plant is used by the Goddess Flora to make his
mother, Juno, conceive. As digitalis is considered scared to both

Flora and Juno, it is thought it was foxglove bells Flora slipped onto her thumb before invoking her powers

Holy Juno grieved that Jupiter had not needed her
services when Minerva was born without a mother.
She went to complain of her husband's doings to Ocean;
tired by the journey, she halted at my [Flora's] door.
...Straightaway I plucked with my thumb the clinging flower and touched Juno,
And she conceived when it touched her bosom.

Book V, 210–263
*Fasti* by Publius Ovidius Naso,[48] 8th Century AD

Foxglove was also thought to ward off witches and a black dye made from the flowers was used in Wales to paint crosses on stone cottage walls and floors.[33] When foxglove stalks bend in the wind, it's believed they are acknowledging a passing supernatural presence.[49]

## 6.5    MODERN MEDICAL USES

As we have seen, digoxin is usually prescribed for heart problems and in emergencies.[50] It is normally safe to use, but toxicity from prescribed digoxin can occur in very old or very young patients.[51] In homeopathy, digitalis is used for cardiac insufficiency and migraine.[6] Digitoxin is used to treat congestive heart failure.[52]

Foxglove leaf (*Mao Di Huang*) is used in Traditional Chinese Medicine (TCM) to strengthen the heart in those with cardiac failure and to encourage urinating in those with edema (swelling) due to heart disease.[53] As well as foxglove, some ginseng plants used in TCM contain chemicals such as **eleutheroside**, which are closely related to digoxin. There was a case reported where a patient prescribed digoxin also took Siberian ginseng as a food supplement, and appeared to have raised serum digoxin levels.[54,55] Other plants can also resemble digoxin in the body. One example is pong-pong seeds (from *Cerbera odollam*, also called the "suicide tree", Figure 6.3) which contain the digoxin-like chemical **cerberin**.[56,57] Although pong-pong seeds do not contain digoxin themselves, patients in A&E who have consumed them often appear to be positive for digoxin.[58–60]

**Figure 6.3**   The Cerbera odollam plant. Reproduced from https://commons. wikimedia.org/wiki/File:Cerbera_odollam_(Flowers).JPG, under the terms of the CC BY-SA 3.0 license, https://creativecommons. org/licenses/by-sa/3.0/deed.en.

**Figure 6.4**   A cane toad (*Bufo marinus*).

Another, more unusual case from 2003 in the USA involved a 40-year-old man taking TCM aphrodisiac pills made from dried cane toad (Figure 6.4) skin. At the hospital he showed all the signs of digitalis poisoning and was positive for digoxin, despite cane toads containing a different cardiac glycoside – **bufadienolide**.[61] Some TCM pills known as *Chan Su* (brand name Kyushin®) intended to treat heart failure are manufactured in the Republic of Korea and also contain poison from a toad. In 2020 an 81-year-old woman who lived in a nursing home showed signs of digitalis poisoning after accidentally taking up to 100 pills. She was treated in hospital with CPR and six lots of defibrillation before recovering. Although she had not taken digitalis, her laboratory tests were positive for both digoxin and digitoxin.[62]

The explanation behind these false positives is likely chemical similarity. As mentioned earlier, there are rapid tests that can be done by hospitals and also in forensic toxicology laboratories as an initial search of a sample for drugs and poisons (a "screen"). One of these tests, known as **immunoassay**, is based on antibody–antigen binding (see Section 6.2). The test is set up so that when the antibody and antigen meet and connect, a chemical or physical change happens that the tester can see or measure – for example a change in colour or **fluorescence**. The intensity of the light is measured and converted to a number that relates to the level of antigen (in this case digoxin) in the sample. Whilst this sounds like it would be very specific, in fact only some of the antibody is available to pair up with the antigen – the rest is attached to the colourful or glowing molecule needed to 'read' the test. The part of the antibody that *is* visible will draw in other similar drugs or poisons and "cross-react" with them, giving a false positive or falsely high reading.[63] It also means that even if digoxin is present in the sample, if other similar compounds are also there (including its metabolites),[52] the level of digoxin measured by the immunoassay is not particularly useful.[14] This can cause confusion in the laboratory and result in incorrect treatment for the patient.[64] The problem of cross-reactivity is the reason immunoassays can't be used as evidence in Court. A presumptively positive immunoassay result must be confirmed by a more specific and sensitive technique such as **GC-MS** (see Chapter 3) or **LC-MS** (see Chapter 2).

Another explanation could be that chemically similar compounds are turned into digoxin in the body (including its fellow alkaloid digitoxin).[52,55] This is a known problem in forensic toxicology and can mean that someone taking a legitimately prescribed medicine ends up testing positive for an illegal drug – for example the Parkinson's drug **selegiline** is transformed into **methamphetamine** and **amphetamine** in the body.[65]

### 6.6  MODERN TOXICOLOGY CASES

Because digitalis has such a profound effect on the heart, if a patient takes it alongside any other heart medications *e.g.*, **quinidine** this can increase the risk of cardiac arrhythmias.[6]

Digoxin is also one of the worst culprits for interacting with herbal medicines such as aloe vera.[64] We have seen already how important potassium is – and aloe vera and senna are known to reduce potassium levels, making digoxin even more toxic. Hawthorn or *Crataegus* (also taken for heart problems[66]) was thought to increase serum digoxin concentrations, but this was revealed to be the old cross-reactivity problem of **epicatechin** from hawthorn giving a false positive on the immunoassay test for digoxin.[67] However, a real change in serum levels of digoxin is seen when patients also take St John's Wort. This time it's a decrease, which is caused by St John's Wort sending one of the enzymes responsible for breaking digoxin down in the body into overdrive.[68]

Digitalis can also be misused for recreation, as the staff of the Poisons Information Centre in Austria discovered between 2002 and 2017. The most likely group to be trying plants for pleasure were youths aged 15 to 20 years.[69] Some of the cases dealt with by the Austrian team were also deliberate suicide attempts, with seven non-fatal ingestions involving digitalis during the same period.[69]

Poisonings in children also happen, and in New Zealand between 2003 and 2010 the Poison Centre received 61 calls about foxglove poisoning, over 70% of which concerned children.[7] Most of the cases involved ingestion of plant material (flowers or leaves). Interestingly, children seem to be able to tolerate higher concentrations of digoxin than adults.[70]

In Canada in 2016, a 67-year-old Chinese woman went to A&E after three days of nausea, vomiting, palpitations and slow and irregular pulse. A few days later her husband joined her in hospital with similar symptoms, both claiming to see a yellow halo around objects after eating some 'kale' from the garden. After being treated with the antibody, both made a full recovery.[14]

Normal therapeutic concentrations of prescribed digoxin are $0.0005–0.0012$ mg $L^{-1}$ in serum.[5,70,71] In a case in 2018 involving an 83-year-old woman in the USA who accidentally took too many of her digoxin pills, her serum level was 0.009 mg $L^{-1}$.[72] In France in 2020, a 71-year-old woman accidentally put some foxglove leaves in her fruit and vegetable juicer. In hospital she had an irregular heartbeat, was vomiting and needed 15 vials of DigiFab® over 6 days to recover.[5] Twenty hours after drinking the juice, her digoxin level was 0.00412 mg L$^{-1}$.[5] Another adventurous juicer, this time in Ireland in 2021, added some unknown leaves to his lettuce and cucumber blend. This 22-year-old man was vomiting, drowsy, light-headed and had blurred vision.[73] In his case, his digoxin level was 0.0006 mg L$^{-1}$, which is within the normal range. However, he was still suffering from poisoning symptoms. To explain this, we need to consider **tolerance** (see Chapter 2); someone who is taking digoxin every day for a heart problem will respond differently to that level in their serum than a naïve user.[14] Secondly, in the case of a plant poisoning, there will be other chemicals present *e.g.*, digitoxin that are not picked up by testing, but may act synergistically with digoxin (see Chapter 3) to cause the symptoms.[5] In another case in 2019 in Italy, a 55-year-old woman was hospitalised after eating a pie made with foxglove leaves she'd mistaken for comfrey.[74] Her digoxin level was 0.01 mg L$^{-1}$, 10 times the top of the therapeutic range, but her symptoms were mild. The explanation for this case could be the cross-reactivity with the **immunoassay** test mentioned earlier to give a falsely high reading. Another complication with digoxin, is that symptoms also depend on the potassium level – a patient could have a normal digoxin level but a life-threatening amount of potassium in their blood[72] or *vice versa*.

The cases we have looked at so far have all been non-fatal, so the levels measured were in serum samples (see Chapter 5) taken at the hospital. In a fatality, we would be measuring the amount of digoxin in post-mortem whole blood and these levels are very challenging to interpret; in Chapters 3 and 4 we met the concept of drug redistribution, and learned how it can cause problems for forensic toxicologists. Post-mortem redistribution wasn't understood until the 1970s, and digoxin was one of the first drugs to be investigated.[75,76] An alternative post-mortem specimen to blood is the fluid from inside the eye, known as "vitreous humour".[76] This clear liquid needs to be sucked out from behind the iris with a needle, and then replaced with saline to re-inflate the eye afterwards.[77] It can be difficult to match up vitreous levels to blood levels, but for digoxin, vitreous levels are not affected by redistribution for 24 h after death.[52] Of course, this is only useful if we know exactly when the person died – for those in hospital the time will be recorded, but for people found deceased at home it can sometimes be a guessing game.

Although foxglove plants grow freely in gardens and parks, prepared digitalis leaves (as a herbal medicine ingredient) are a prescription-only medicine in the UK and can only be made available by a registered doctor.[78]

### 6.7 CATCHING THE CHEMICAL CULPRIT

Digoxin and digitoxin can be extracted from blood in toxicology cases using a separating technique known as **liquid–liquid extraction** (**LLE**), which was discussed in Chapter 4.[79] As we saw earlier, the standard hospital method of **immunoassay** can be fooled by similar chemicals from other plants, so a more specific method such as **LC-MS** is needed if results are to be used in Court.[79]

### 6.8 CASE CLOSED

Naturally, we expect high drama from our Bond films, but after taking digoxin orally, the effects probably wouldn't appear for about 2 h,[27] never mind the 2 min we see in *Casino Royale*. Although once they do appear, Bond's symptoms of nausea, sweating and heart problems are what we might expect from a digitalis overdose. However, the heart monitor Bond is hooked up to in the film shows his heart racing at 136 bpm (normal resting heart rate for men is 60–100 bpm), which as we have seen is the opposite to a real digitalis poisoning, where the heart slows to a stop.[80]

Even less convincing was his treatment and recovery. As we saw earlier, in real-life poisonings, patients can end up in intensive care for days, needing multiple vials of the expensive antidote DigiFab® – although we can assume that cost would be no barrier to MI6 for their most famous secret agent.

### REFERENCES

1. N. Purvis and R. Wade, *Casino Royale Screenplay*, Danjac LLC, Santa Monica, CA, 2005.
2. L. Gautam, S. D. Sharratt and M. D. Cole, *PLoS One*, 2014, **9**, e89031.
3. K. Harkup, *Chem. World*, 2020, **17**, 54–57.
4. Cardiac arrest, https://www.bhf.org.uk/informationsupport/conditions/cardiac-arrest, accessed May 2021.

5. E. Rouault, C. Ghnassia, E. Filippi-Codaccioni and N. Maillard, *Basic Clin. Pharmacol. Toxicol.*, 2021, **128**, 183–186.

6. Medical Economics Company, *PDR for Herbal Medicines*, Medical Economics Company, Montvale, NJ, 1998.

7. R. J. Slaughter, M. G. Beasley, B. S. Lambie, G. T. Wilkins and L. J. Schep, *N. Z. Med. J.*, 2012, **125**, 87–118.

8. A. Chevallier, *Encyclopedia of Medicinal Plants*, DK Publishing, St Leonards, 2001.

9. A. Stewart, *Wicked Plants*, Algonquin Books of Chapel Hill, Chapel Hill, NC, 2009.

10. J. Emsley, *More Molecules of Murder*, RSC Publishing, Cambridge, 2017.

11. M. Brown, *Death in the Garden: Poisonous Plants & Their Use Throughout History*, Pen & Sword Books Ltd, Barnsley, 2018.

12. D. Frohne and H. J. Pfänder, *Poisonous Plants: A Handbook for Doctors, Pharmacists, Toxicologists, Biologists and Veterinarians*, Manson Publishing, London, 2nd edn, 2004.

13. C. C. Lin, C. C. Yang, D.-H. Phua, J.-F. Deng and L.-H. Lu, *J. Chin. Med. Assoc.*, 2010, **73**, 97–100.

14. R. M. Janssen, M. Berg and D. H. Ovakim, *Can. Med. Assoc. J.*, 2016, **188**, 747–750.

15. J. H. Bock and D. O. Norris, in *Forensic Plant Science*, ed. J. H. Bock and D. O. Norris, Academic Press, San Diego, 2016, ch. 1, pp. 1–22.

16. *Gerard's Herbal*, ed. M. Woodward, Senate, London, 1994.

17. E. Stoye, in *Chemistry in its Element*, RSC Publishing, Cambridge, 2013.

18. A. Foster, *The Medicinal Plant Collection at the University of Oxford Botanic Garden*, Wellcome Trust, Oxford, 2010.

19. M. Levine, A.-M. Ruha, K. Graeme, D. E. Brooks, J. Canning and S. C. Curry, *Chest*, 2011, **140**, 1357–1370.

20. Anonymous, *The Poison Garden*, The Alnwick Garden, Alnwick, 2005.

21. C. Hogue, *Chem. Eng. News*, 2005, **83**, 58.

22. M. R. Cooper, A. W. Johnson and E. A. Dauncey, *Poisonous Plants and Fungi: An Illustrated Guide*, The Stationery Office, Norwich, 2nd edn, 2003.

23. Medicines from Plants, https://s3-eu-west-1.amazonaws.com/assets.botanic.cam.ac.uk/wp-content/uploads/2019/12/medicines_trail_web.pdf, accessed July 2021.

24. N. Bailey, *Chelsea Physic Garden: Connecting People with Plants since 1673*, Chelsea Physic Garden, London, 2015.
25. W. Withering, *An Account of the Foxglove and Some of its Medical Uses with Practical Remarks on Dropsy*, G. G. J. and J. Robinson, London, 1785.
26. The Royal Botanical Gardens Kew, *Plants+People*, Kew Publishing, London, 1998.
27. J. Patocka, E. Nepovimova, W. Wu and K. Kuca, *Environ. Toxicol. Pharmacol.*, 2020, **79**, 103400.
28. M. Grieve, *A Modern Herbal*, Dover Publications, New York, 2nd edn, 1971.
29. R. Bevan-Jones, *Poisonous Plants: A Cultural and Social History*, Windgather Press, Oxford, 2009.
30. G. Eliot, *Silas Marner*, Wordsworth Editions Limited, Ware, 1999.
31. H. B. Burchell, *J. Am. Coll. Cardiol.*, 1983, **1**, 506–516.
32. M. Webb, *Precious Bane*, Oberon Books, London, 2003.
33. F. Inkwright, *Botanical Curses and Poisons*, Liminal 11, London, 2021.
34. W. Wordsworth, *Wordsworth Poetical Works*, Oxford University Press, Oxford, 1984.
35. C. R. Mackin, J. F. Peña, M. A. Blanco, N. J. Balfour and M. C. Castellanos, *J. Ecol.*, 2021, **109**, 2234–2246.
36. A. Tennyson, *Tennyson: In Memoriam*, Clarendon Press, Oxford, 1984.
37. W. Shakespeare, *Hamlet*, Oxford University Press, Oxford, 1998.
38. K. Harkup, *Death by Shakespeare: Snakebites, Stabbings and Broken Hearts*, Bloomsbury, London, 2020.
39. M. H. A. Piette and E. A. De Letter, *Forensic Sci. Int.*, 2006, **163**, 1–9.
40. D. Frohne and H. J. Pfänder, *A Colour Atlas of Poisonous Plants*, Wolfe Publishing Ltd, London, 1983.
41. M. Willes, *A Shakespearean Botanical*, Bodleian Library, Oxford, 2020.
42. S. T. Dietz, *The Complete Language of Flowers: A Definitive and Illustrated History*, Wellfleet Press, New York, 2020.
43. Ryn, in *Botany After Dark*, Podtail, Stockholm, 2021.
44. W. Turner, *A New Herball*, Cambridge University Press, Cambridge, 1989.
45. F. Inkwright, *Folk Magic and Healing: An Unusual History of Everyday Plants*, Liminal 11, London, 2019.

46. S. Lawrence, *Witch's Garden: Plants in Folklore, Magic and Traditional Medicine*, Welbeck, London, 2020.
47. R. Vickery, *Garlands, Conckers and Mother-Die: British and Irish Plant-Lore*, Continuum, London, 2010.
48. Ovid, *Fasti*, Harvard University Press, Cambridge, MA, 2nd edn, 1989.
49. R. Richardson, *Britain's Wild Flowers: A Treasury of Traditions, Superstitions, Remedies and Literature*, The National Trust, London, 2017.
50. Royal Pharmaceutical Society, *British National Formulary*, BNF Publications, London, 2021.
51. L. S. Nelson, M. A. Howland, N. A. Lewin, S. W. Smith, L. R. Goldfrank and R. S. Hoffman, *Goldfrank's Toxicologic Emergencies*, McGraw Hill, New York, 11th edn, 2018.
52. *Disposition of Toxic Drugs and Chemicals in Man*, ed. R. C. Baselt, Biomedical Publications, Seal Beach, 12th edn, 2020.
53. J. Zhou, G. Xie and X. Yan, *Encyclopedia of Traditional Chinese Medicines*, Springer, Heidelberg, 2011.
54. N. K. Ho, *Singapore Med. J.*, 2001, **42**, 487–492.
55. S. McRae, *Can. Med. Assoc. J.*, 1996, **155**, 293–295.
56. J. M. Rague, L. S. Halmo and K. Heard, *American College of Medical Toxicology 2020 Annual Scientific Meeting New York*, 2020, p. 163.
57. Y. Gaillard, A. Krishnamoorthy and F. Bevalot, *J. Ethnopharmacol.*, 2004, **95**, 123–126.
58. H. Fok, P. Victor, S. Bradberry and M. Eddleston, *Clin. Toxicol.*, 2018, **56**, 304–306.
59. A. Holzer, S. Dorner-Schulmeister, K. Bartecka-Mino and D. Genser, *40th International Congress of the European Association of Poisons Centres and Clinical Toxicologists (EAPCCT)*, Tallinn, Estonia, 2020, p. 536.
60. S. P. Nordt, M. Hendrickson, K. Won, M. J. Miller, S. P. Swadron and F. L. Cantrell, *Am. J. Emerg. Med.*, 2020, **38**, 1698. e1695–1698.e1696.
61. R. M. Gowda, R. A. Cohen and I. A. Khan, *Heart*, 2003, **89**, e14.
62. K. Cha, B. H. So and W. J. Jeong, *Hong Kong J. Emerg. Med.*, 2020, **27**, 180–184.
63. M. E. Wermuth, R. Vohra, N. Bowman, R. B. Furbee and D. E. Rusyniak, *J. Emerg. Med.*, 2018, **55**, 507–511.
64. H. H. Tsai, H. W. Lin, A. Simon Pickard, H. Y. Tsai and G. B. Mahady, *Int. J. Clin. Pract.*, 2012, **66**, 1056–1078.

65. F. Karoum, L.-W. Chuang, T. Eisler, D. B. Calne, M. R. Liebowitz, F. M. Quitkin, D. F. Klein and R. J. Wyatt, *Neurology*, 1982, **32**, 503–509.
66. J. Barnes, *Herbal Medicines*, Pharmaceutical Press, London, 2013.
67. A. Dasgupta, L. Kidd, B. J. Poindexter and R. J. Bick, *Arch. Pathol. Lab. Med.*, 2010, **134**, 1188–1192.
68. A. Johne, J. Brockmöller, S. Bauer, A. Maurer, M. Langheinrich and I. Roots, *Clin. Pharmacol. Ther.*, 1999, **66**, 338–345.
69. S. Dorner-Schulmeister, K. Bartecka-Mino and A. Holzer, *40th International Congress of the European Association of Poisons Centres and Clinical Toxicologists (EAPCCT)*, Tallinn, Estonia, 2020, p. 537.
70. *Clarke's Analysis of Drugs and Poisons*, ed. A. C. Moffatt, D. Osselton and B. Widdop, Pharmaceutical Press, London, 4th edn, 2011.
71. M. Schulz, A. Schmoldt, H. Andresen-Streichert and S. Iwersen-Bergmann, *Crit. Care*, 2020, **24**, 195.
72. M. L. P. Mattison, V. V. Muse, L. H. Simmons, C. Newton-Cheh and R. K. Crotty, *N. Engl. J. Med.*, 2018, **378**, 1931–1938.
73. T. Popoola, E. Umana and J. Binchy, *Ir. Med. J.*, 2021, **114**, 245.
74. M. S. Negroni, A. Marengo, D. Caruso, A. Tayar, P. Rubiolo, F. Giavarini, S. Persampieri, E. Sangiovanni, F. Davanzo, S. Carugo, M. L. Colombo and M. Dell'Agli, *Case Rep. Cardiol.*, 2019, **2019**, 9707428.
75. D. W. Holt and J. G. Benstead, *J. Clin. Pathol.*, 1975, **28**, 483–486.
76. R. J. Dinis-Oliveira, F. Carvalho, J. A. Duarte, F. Remião, A. Marques, A. Santos and T. Magalhães, *Toxicol. Mech. Methods*, 2010, **20**, 363–414.
77. C. Valentine, *Murder Isn't Easy: The Forensics of Agatha Christie*, Sphere, London, 2021.
78. Banned and restricted herbal ingredients, https://www.gov.uk/government/publications/list-of-banned-or-restricted-herbal-ingredients-for-medicinal-use/banned-and-restricted-herbal-ingredients, accessed May 2020.
79. E. L. Øiestad, U. Johansen, M. S. Opdal, S. Bergan and A. S. Christophersen, *J. Anal. Toxicol.*, 2009, **33**, 372–378.
80. K. Harkup, *Superspy Science: Science, Death and Tech in the World of James Bond*, Bloomsbury, London, 2022.

# CHAPTER 7

# Poison for a Broken Heart

If you see a term that's **bold** it's defined in the Glossary. Only the first time that the word appears in the chapter will it be indicated in this way.

---

**Case History: Crime of Passion**

At night she began cooking things in the kitchen, things too strange to mention. She steeped oleander in boiling water, and the roots of a vine with white trumpet flowers that glowed like faces. She soaked a plant collected in moonlight from the neighbors' fence, with little heart-shaped flowers. Then she cooked the water down; the whole kitchen smelled like green and rotting leaves.

p. 36

*White Oleander* by Janet Fitch,[1] 1999

---

## 7.1 THE INVESTIGATION

In the case history, 12-year-old Astrid describes how her mother, Ingrid, boiled up several plants to make what we later find out to be a poisonous infusion. She uses it on her ex-lover Barry who has jilted her for another woman, but there are few details of the poisoning itself in *White Oleander*, so we do not know the effects

---

Poisonous Tales: A Forensic Examination of Poisons in Fiction
By Hilary Hamnett
© Hilary Hamnett 2023
Published by the Royal Society of Chemistry, www.rsc.org

experienced by him. We learn later in the book that Ingrid added **dimethyl sulfoxide (DMSO)** to the concoction and painted it on his doorknob, and that the poisoning was fatal. Ingrid is found guilty of murder and sent to prison for life, leaving Astrid to navigate a difficult journey through various foster homes.

Smearing poison on a doorknob might have been simply an intriguing plot twist back in 1999, but became a stark reality in 2018 with the poisoning of Sergei and Yulia Skripal in Salisbury in the UK. Both were poisoned by a novichok nerve agent painted on their doorknob and spent several weeks in hospital. Forensic scientists at the scene found the agent in several places, but the highest concentration was on their front door.[2]

The scientists who attend crime scenes are specialist examiners, trained to look for all different types of evidence (not just things that are toxicologically significant). A forensic toxicologist enters later in the case when the samples taken at the scene or autopsy are delivered to the lab. This means that if the samples are not useful (because they're the wrong kind or there's not enough) it's too late to do anything about it. Occasionally a toxicologist will be asked for advice before the samples are taken, but not often. Forensic toxicologists are also rarely called upon to analyse samples swabbed at crime scenes, or powders, pills or plants seized from suspects or found by customs officers. These samples are usually sent to forensic drugs laboratories, where similar analytical techniques are used to identify them.[3] We will see in Chapter 9 that there are some nifty quick forensic chemistry tests that can be done at crime scenes to help guide the investigation later on.

In our case, Ingrid left the police several clues by breaking into Barry's house a few weeks earlier

This time she put a sprig of oleander in his milk, another in his oyster sauce, in his cottage cheese. She stuck one in his tooth-paste. She made an arrangement of white oleanders in a hand-blown vase on his coffee table, and scattered blooms on his bed.

p. 32
*White Oleander* by Janet Fitch,[1] 1999

What's more, the white oleanders in question were actually growing outside her house. These were not the only plants she used in her poisoning, and we will explore the others in later sections.

Dimethyl sulfoxide is a by-product of the paper-making process and a chemical that is used to dissolve other chemicals (a "solvent") to make them easier to use. Although it now has its own medical uses, DMSO was originally used as a carrier. This meant medications were dissolved in it then applied to the skin, usually by rubbing or dabbing directly with the hands.

## 7.2 THE PLANTS BEHIND THE STORY

### 7.2.1 Oleander

Common oleander (*Nerium oleander*) is also known as "adelfa", "rose laurel", "rose bay", "rosa Francesca" and "laurier rose".[4] It has many slender stems near ground level and dark dull-green pointy leaves.[5] Although native to the Mediterranean, these shrub-like plants grow outside in warmer parts of the UK or in glasshouses. Outside, oleanders grow in enclosed greens, sea-bordering places and near rivers.[6,7] Although we know the oleanders in our case were white (Figure 7.1, left), the large clusters of flowers at the tips of twigs can also be red or pink (Figure 7.1,

**Figure 7.1** Common oleander plants with (left) white flowers, taken at the Oxford Botanic Garden and Arboretum and (right) pink flowers, taken in Lincoln.

right),[8,9] making them popular pot plants.[10] White oleanders look similar to the *Cerbera odollam* plant (or "suicide tree") we met in Chapter 6, and have also been mistaken for olive.[7] Yellow oleanders are a different species called *Thevetia peruviana* (Figure 7.2).

Oleanders flower from May to August, and similar to opium poppies (see Chapter 2) secrete a milky juice containing the active chemicals.[11] All parts of the plant are poisonous, and unlike some of the previous plants we have encountered, drying does not affect the potency, and its dry leaves are as toxic as its fresh ones.[12]

Earlier in the book, Ingrid implies that she plans to poison Barry with oleander by mouth, leaving a warning when she puts whole oleander flowers in food and drink around his house. But this poisoning plan would not have been very successful; oleander is actually unpalatable because of its bitter and pungent taste,[13] so Barry would probably have guessed something was up before he took a fatal dose. The taste does not deter everyone though; drinking a tea made from oleander leaves is a known suicide method, and children have also become sick after eating a few leaves or petals (attracted by the enticing flowers).[10]

Every part of the plant contains **cardiac glycosides** (see Chapter 6), chemicals that affect the heart, including **oleandrin**, **neriine**, and **digitoxigenin**.[8,14] Collectively these compounds are referred to as "cardenolides" and the most active molecule is oleandrin.[13] They are also responsible for its unpleasant taste. The highest concentration of cardenolides in both common and

**Figure 7.2**    Yellow oleander (*Thevetia peruviana*).

yellow oleander can be found in the seeds,[10] but the yellow oleander contains other compounds such as **thevetin A**.[15] *Nerium oleander* plants with red flowers seem to be more poisonous than those with white flowers.[9]

Oleander poisoning usually starts within a few hours of ingestion with a burning sensation in the mouth, and a bitter taste that leads to abdominal pain, and very quickly vomiting and diarrhoea.[16] Like digitalis in Chapter 6, victims sometimes see a yellow or green halo and have a headache.[7] The most toxic effects of oleandrin are on the heart, and death is due to a slow, weak and irregular pulse followed by a big drop in blood pressure.[8] Most deaths happen within 24 h, but the toxicity of oleander seeds depends on how well they are crushed (so some deaths may take longer).[9] Because the effects such as cardiac arrhythmias and electrolyte disturbances (*e.g.*, changes in potassium concentrations), are so similar to **digoxin** or yellow oleander poisoning, sometimes A&E doctors mistake the source of the symptoms.[17]

A favourite in folk medicine for centuries, and associated with St Joseph,[18] you can still buy capsules containing 400 mg of oleander leaf extract in the USA as a dietary supplement to treat a wide variety of physical ailments.[19] Such things as obesity, erectile dysfunction, herpes and eczema can apparently be cured.[16]

In the past it was mainly used to treat disorders of the heart (rather than revenge after unhappy love affairs), as well as skin diseases such as leprosy.[20] Gerard noted how dangerous oleander was to humans and animals, but he and Turner still suggested drinking oleander mixed with wine as a cure for the sting or bite of a serpent (see Chapter 9).[21,22]

When collected for medicinal use, the leaves are picked shortly before flowering and then dried in the shade[7] (the oleandrin concentration is highest at the point of flowering[23]). Although the danger remains even after drying or boiling.[8] Like many herbal products, oleander supplements are often misrepresented as 'harmless' because they are natural products. Given the similarity with cardiac glycosides, you may be wondering if oleanders were ever explored as heart medications. In the early 1930s they were, but the gastrointestinal side-effects were too severe.[9]

Oleander acts by interfering with the movement of sodium and potassium around the cells of the heart, in a similar way to digitalis (Chapter 6),[13] but is generally weaker, probably due to the lower rate of absorption of oleandrin.[7]

### 7.2.2 Datura

In the novel, Ingrid also added "white trumpet flowers" to her infusion, and these could be a few different plants. The most common one fitting this description in the UK is bindweed (*Calystegia sepium*), which can be seen flowering in late summer. Although bindweed *is* deadly, it kills other plants by winding around and smothering them, it is not actually poisonous. That leaves us with Angel's trumpet (*Brugmansia*), which we met in Chapter 2, or Devil's trumpet (*Datura stramonium*, Figure 7.3), both of which contain **scopolamine** (see Chapter 2),[24] as the most likely candidates responsible for Barry's demise. The hanging flowers of *Brugmansia* (Figure 2.3), are said to blast from the Heavens down towards the Devil, whereas those of Datura are said to call upwards to the Heavens.

Datura is also known as "thorn apple" (because of its spikey seed pod that looks a bit like a horse chestnut,[25] Figure 7.3,

**Figure 7.3**    (Left) A *Datura stramonium* plant in bloom, taken at the Chelsea Physic Garden. (Right) A Datura seed pod, taken at the Oxford Botanic Garden and Arboretum.

right), "jimsonweed", "locoweed" or "Devil's snare" (familiar to Harry Potter fans), and it grows on rubbish tips, in crop fields and on waysides.[26] The flowers are most likely to be white, but can also be purple.[27] Although we now think of it as a pesky annual weed,[28] it's not native to the UK and today's plants are all descended from (or 'escapees' from) deliberately introduced plants.[29] Its leaves are jagged and can be as long as 20 cm, and the flowers are sweet-smelling long tubes that splay at the end like a trumpet, and can be white or purple, but close at night.[8,30-32] Unlike many of the plants we have met so far, it is the flowers, appearing May to September, which contain the highest amount of alkaloids.[10,11] Poisonings with the leaves and seeds can still happen when they are mistaken for other plants such as nettle, chilli (capsicum) or when deliberately smoked or ingested for recreational reasons.[10] The name "stinkweed" comes from the plant's disagreeable odour.[8,33] Fortunately, most of the plant has a nauseous taste, but the sweeter ripe seeds may attract children.[33] Gerard tells us that despite its offensive odour it causes drowsiness.[34]

Datura poisoning closely resembles overdose with digitalis (Chapter 6) but particularly affects the eyes. Starting with dimness of sight, dilation of the pupils and progressing to blurred vision, the eyes can be affected even if you just accidentally rub them after handling the plant. Other symptoms such as gastro-intestinal problems, headache, dizziness and heart arrhythmias follow.[35] Signs and symptoms usually start to show within half an hour but can continue for up to two days, because Datura slows down its own passage through the bowels.[4] Psychoactive symptoms are common with Datura such as hallucinations, overtalkativeness, convulsive sobbing and sexual excitement, as well as aggressive behaviour.[36]

Of all the plants we have encountered so far, Datura is the one most likely to be used recreationally, because of its mind-altering properties. These have been known for centuries, and tale has it that some British soldiers in Jamestown in the USA in 1676 ate some Datura leaves in a salad and experienced the hallucinogenic effects, giving it the name "jimsonweed". Versions of the story vary, with some reporting it was an accident and others claiming the colonists set out to poison the British who had come to quash a local uprising.[32] The fresh plant is rarely eaten, probably because of the unpleasant odour, but teas made from dry leaves, cigarettes, and the seeds have been consumed since the 1800s for the powerful visions, sounds and feelings they conjure up.[28] Like any hallucinogen, unfortunately these pleasant sensations can turn into paranoia, anxiety, panic and confusion. Forensic toxicologists will see accidental deaths of people under the influence of hallucinogens, such as road traffic crashes (fleeing from phantom pursuers) or falls from height (thinking they can fly).

In Europe there are also some folk medicine uses for Datura; the leaves can be smoked in a cigarette (Brosig's cigarettes)[37] or a pipe to treat asthma and reduce pain and swelling,[28,33,38] as burning does not deactivate the alkaloids.[39] A tincture made from the fruit has also been used for diarrhoea and it has even been tried as a cure for alcoholism.[10]

## 7.3 POISONOUS PLOTS

### 7.3.1 Datura

Poisoning with a Datura leaf is key to the plot of Leo Delibes' opera *Lakmé*, which premiered in 1883. A somewhat obscure piece, it was made famous by the use of the Act I song "Viens, Mallika" or "Flower Duet" by *British Airways* in its advertising.[40] Set during colonial rule in India, Lakmé the daughter of a Hindu high priest, falls in love with a British soldier named Gerald. Her father does not approve and the couple run away together to a cabin in the forest. When Lakmé leaves the cabin in search of mystic water (that gives everlasting love to the drinkers), Gerald is visited and persuaded to put duty first by a fellow soldier. When she returns and realises he is going to desert her, rather than live with the dishonour of what they have done, she takes a bite from a Datura leaf[†] and dies[41]

| LAKME: | LAKME: |
|---|---|
| Tu n'oses pas. | Thou darest not drink! |
| *Elle regarde attentivement Gérald* | *(Looks attentively at Gerald...)* |
| C'est là-bas que va sa pensée... | Yonder, ah! His thought are flying. |

---

[†]Note the mis-translation of the French 'feuille' as 'flower', when it actually means 'leaf'. Although as we have seen, the flowers are the most poisonous parts of the plant, so this does not affect the plot of the opera.

| | |
|---|---|
| Son cœur a tressailli, | His heart is throbbing fast, |
| Et sa patrie à ses yeux s'est dressée. | And for his home far away he is sighing! |
| *Avec déchirement après avoir essayé d'attirer son regard.* | *(In despair after having tried to attract his gaze.)* |
| Tout est fini! | All's over now! |
| *...Lakmé, désespéree, arache une feuille de datura et la mâche sans que Gérald s'en aperçoive.* | *(...Lakmé plucks a datura flower and bites it in twain, smiling and unseen by Gerald.)* |

p. 45
*Lakmé* by Leo Delibes (ed. Burden),[42] 1883

It's questionable whether one leaf (or flower) could contain enough alkaloid to be fatal, but as we will see, eating, drinking or smoking substantial amounts of the plant can cause serious toxicity.[4]

The novel *Madapple* by Christina Meldrum published in 2008, takes its title from another name for Datura.[43] With some similar themes to *White Oleander*, a daughter (Aslaug) brought up in isolation is left to fend for a herself after her mother (Maren) dies suddenly. In her childhood they had frequently foraged for medicinal and edible plants

Mother crouches in the field, her body folded into itself as she
uproots the salsify plants and lops off their purple heads....
Yet I know it is the jimsonweed, not the nettle she is prepared to find:
Rank-smelling, rash causing, poisonous jimsonweed.
When I was growing up, Mother called the weed a variety of names.
Mad apple at times, Devil's apple at times.

p. 7
*Madapple* by Christina Meldrum,[44] 2008

Aslaug reveals later that her mother was secretly burning the dried jimsonweed leaves and seeds and inhaling the smoke for its hallucinogenic effects.

Datura also features in the supernatural thriller *The Autopsy of Jane Doe* where it is found in the stomach of an unidentified young woman, who it turns out was falsely accused of witchcraft.

### 7.3.2 Oleander

An interesting urban legend that circulates in the USA and Australia, involves a troop of Boy Scouts being fatally poisoned when they accidentally used oleander sticks to roast treats over a campfire. Reports of the poisoning vary, with the roasted food being marshmallows in some and hot dogs in others. In fact, Astrid has heard this story in *White Oleander*

I stood on the porch and gazed at the giant oleander.
It was old, it had a trunk like a tree.
You just had to roast a marshmallow on one twig and you were dead.
She'd boiled pounds of it to make the brew of Barry's death.
I wondered why it had to be so poisonous.

p. 361
*White Oleander* by Janet Fitch,[1] 1999

This legend continued until 2021 when an inspired medical researcher roasted their own hotdogs skewered on oleander sticks over a disposable BBQ and tested the oleandrin content in the meat using **LC-MS** (see Chapter 2). The amount was so small that the author estimated the Boy Scouts would have needed to eat over 180 hot dogs each to feel any effects.[45]

*Dragonwyck* by Anya Seton (1944) is a gothic novel that tells the story of a couple Johanna and Nicholas Van Ryn who live in Dragonwyck mansion in the USA in the 1840s. Johanna is suffering from a cold but takes a turn for the worse and dies suddenly

There was a horrible noise in the room, the sound of retching, steady, almost rhythmic. On the bed, a shapeless figure threw itself backward and forward in a monotonous, mechanical way.
Jeff [the doctor] stood for a moment appalled, then he groped for the pulse,
which was terrifyingly slow and irregular; the flesh beneath his fingers was clammy as an eel.

p. 162
*Dragonwyck* by Anya Seton,[46] 1944

It turns out she had eaten rather a lot of a tipsy cake (a fruit sponge soaked in rum or sherry) with 'nutmeg' on the top.

The doctor's suspicions are aroused by her huge pupils and he even slyly collects a piece of cake at the scene to examine, but finds none of the common poisons of the day on it. It is revealed later in the book that Nicholas ground up the leaves of an oleander plant in the room and put them on the cake, the rum disguising the bitter taste. His plan was to kill his wife in order to marry a younger distant relative, Miranda.

The name "oleander" is thought to come from the Greek mythological names of two doomed lovers, Leander and Hero. Leander would swim across a narrow and dangerous strait to visit his lover Hero every night, guided by the light from her tower. One night the light blew out in a storm and he drowned. When she found his body he was clutching [what is now known as] an oleander flower, and she threw herself off her tower in order to join him.[47]

## 7.4 DATURA AND THE DIVINE

Datura has long been used as a sacred hallucinogen in South America and the Himalayas. It was believed that when Buddha preached, dew or raindrops fell from heaven onto Datura.[48] In India it was known as the "god of destruction" and those who drank wine spiked with its seeds lost their memory, seemed oblivious to their surroundings and their senses. Sometimes it was smoked alongside cannabis or tobacco,[48] or used in anaesthesia.[49]

In New Mexico it is used to speak with the dead as it is believed that the plant grows over portals to the realm of their ancestors.[50]

Datura was thought to be an aid to the incantation of witches and considered unlucky for anyone to grow in their garden (although possibly mainly because if found, they would be accused of witchcraft or wizardry).[33]

A more modern take on such a potion was confiscated from a man in Spain in 2009, who was arrested selling an infusion of Datura leaves on the street, labelled "witches brew".[29] Datura is also a contender for Circe's baneful pig-producing potion, which we met in Chapter 2. Jimsonweed still grows all over the classical World, and also produces the amnesia Circe was hoping to see, making her victims forget their homeland.[51]

Native Americans used Datura to break hexes and as part of the rite of passage into adulthood. Sometimes it was the Shaman who would direct the use of Datura for spiritual purposes, or a medicine man would 'prescribe' it to those seeking help for their health problems. In Haiti, Datura is used alongside tetrodotoxin (a highly potent toxin from the puffer fish or *Fugu*[52]) in a ceremony to produce a death-like sleep and dream-like stupor.[50]

## 7.5   OLEANDER AND ORACLES

In another Greek myth, Pythian priestesses were women who acted as the mouthpieces of the god Apollo, issuing prophecies on his behalf. How they communicated with Apollo remains a mystery, but modern sources believe it may have been *via* deliberate oleander intoxication. Ancient texts describe the priestesses chewing laurel leaves and inhaling laurel incense, which would bring on symptoms such as agitation, drooling, loss of senses, and sometimes death. But the most common laurel plant native to the Greeks, the Bay Tree (*Laurus nobilis*, Figure 7.4, left) is not psychoactive.[53] Another possibility is the Cherry Laurel (*Prunus laurocerasus*, Figure 7.4, right), which can give off hydrogen cyanide (HCN) gas when burned (see Chapter 10), and

**Figure 7.4**   A Bay Tree (left), taken at the Turner Garden, and a Cherry Laurel hedge (right), taken in Lincoln.

although death is certainly one of its effects, it's not known as an intoxicant. On the other hand, oleander, which is also known as "rose laurel" grows in the Mediterranean, and its symptoms are a better fit for the activities of the priestesses.[54]

Oleander has also become associated with death – in Tuscany and India the dead are covered in or crowned with oleander before the burial.[50] It is considered unlucky to bring oleander into the home in some countries, and doing so will bring disgrace, misfortune and sickness.[43] In another myth it is believed to be dangerous to sleep in the same room as an oleander.[55]

## 7.6  MODERN MEDICAL USES

### 7.6.1  Oleander

In Indian traditional medicine, oleander ("adelfa") roots and leaves are used to prepare decoctions (green dyes) for various skin conditions, scabies, eye diseases (using only the juice of the leaves) and hemorrhoids.[7,56] In 2012 in Turkey, a 30-year-old male was admitted to hospital after developing a dangerous heart arrhythmia. On the advice of a herbalist he had drunk a homemade syrup of oleander leaves to treat his hemorrhoids. He was discharged after 6 days after his heart was stabilized with the help of a pacemaker.[57]

The root is sometimes used as an abortifacient in rural communities.[4] It is also used as an insect repellant, and in some homeopathic remedies.[58] In traditional medicine, leaves may be applied directly to tumors, but the potential of oleander extract (specifically oleandrin and oleandrigenin) as a chemotherapeutic agent has also been studied,[59] and marketed as the drug Anvirzel™.[60]

Oleander can also be prescribed in Haitian herbal medicine, but unfortunately the dosage isn't always appropriate. A case in Florida in 1987 involved a woman in her 20s being recommended a very strong potion of oleander leaf to be administered orally and rectally. After starting to feel unwell the patient was taken to hospital but died despite attempts to treat her.[20]

In a case in Switzerland in 2004, a 59-year-old man was admitted to hospital with a dangerously low heart rate after using a homemade oleander lotion on his psoriasis. He had picked some flowers and leaves then boiled them twice in water before applying the lotion to his skin. He survived after being fitted with a pacemaker, but carried on dabbling in home treatments, later choosing to remove his skin lesions caused by the psoriasis with a knife.[23]

During 2020, before vaccines against COVID-19 had been approved, some people turned to supplements (most famously vitamin D) for prevention and cure.[61] A clinical trial that ran from May to July 2020 tried a sublingual extract of *Nerium oleander* on patients testing positive for the virus. No results were ever published, but the high-profile individuals taking and promoting it the USA sparked some controversy, which is not surprising given oleander's potentially fatal side-effects.[62]

### 7.6.2 Datura

In folk medicine, Datura preparations have been used for asthma,[29] and coughs during pertussis, bronchitis and influenza. When mixed with hog's grease it was made into a salve for burns and scalds.[31] In homeopathy it is used to treat infection with high temperature, cramps and inflammation of the eyes.[7]

In Traditional Chinese Medicine (TCM) it is used by Tibetan healers for general pain, smoked for asthma, applied externally for rheumatism,[7] and used to treat Parkinson's disease, ankle pain, asthma, cough, gastric convulsions, and traumatic injuries.[63,64] Accidental poisonings with herbal medications made from Datura plants mean there are legal restrictions on buying and selling them. In the UK they can only be sold in premises that are registered pharmacies and by or under the supervision of a pharmacist.[65]

In Ayurvedic medicine, Datura is used to treat various health problems, such as inflammation, ulcers and wounds, where it is applied as a paste or solution to the skin.[66]

### 7.7 MODERN TOXICOLOGY CASES

### 7.7.1 Oleander

Deaths from oleander are usually after voluntary ingestion of decoctions or plant material for suicidal purposes. While 5–15 oleander leaves could cause fatal toxicity to adults, only one might be toxic to a child.[13] Collectively though, there is no consensus between experts about the oleander lethal dose, or even clinical signs of oleander toxicity.[56] Between 2002 and 2017 the Austrian poisons centre received calls about seven non-fatal poisonings involving oleander.[67] Similarly, 9% of the severe plant poisonings handled by intensive care units in France from 2007–2019 involved oleander.[68] Of the 81 calls to the New Zealand poison centre between 2003 and 2010 regarding oleander, over half involved children (53%) accidentally eating the plants.[5]

In 2005 in Germany, a 47-year-old woman was admitted to hospital after eating a bowl of oleanders. She had vomiting and was drowsy, but recovered and was discharged after six days.[69]

As we have seen above, accidental poisonings also occur in those seeking oleander-based herbal treatments. In the United Arab Emirates in 2008, a 49-year-old man died in hospital. He had a complex medical history including diabetes and heart disease, and had apparently been given some leaves by an unknown person to make an infusion to treat his diabetes. The toxicology lab found oleandrin in his blood sample at a concentration of 0.01 mg $L^{-1}$ about 10 times that seen in non-fatal but toxic cases (0.001–0.002 mg $L^{-1}$ [16]).[70] It's possible that the victim's pre-existing medical conditions, known as "co-morbidities", contributed to his death, or meant that a lower concentration of the poison was fatal. This is another factor that makes interpreting forensic toxicology results difficult – if the victim's heart is already under strain from illness, less poison is needed than for an otherwise fit and healthy person.

Oleander tea is still a surprisingly common cause of accidental poisoning, and is dangerous because boiling the leaves does not deactivate the oleandrin.[71] In Italy in 2020, a 71-year-old male was found dead at home after making a drink of oleander tea from a recipe he found on the internet. His oleandrin concentration in post-mortem peripheral blood was 0.0375 mg $L^{-1}$, 40 times the toxic concentration.[16] In 1985 in the USA, a 30-year-old woman was taken to hospital 10 h after making tea out of oleander leaves, mistaking them for eucalyptus. She was confused, vomiting and had a numb tongue. Despite attempts to save her with **atropine** and a pacemaker she died.[71] An 83-year-old resident of a nursing home in the USA was taken to hospital after drinking oleander tea made from a plant she had received by mail order.[39]

As well as accidental poisonings with oleander, there are also deliberate suicide attempts. Seventeen such attempts were reported to the Swiss poisons centre between 1995 and 2009.[72] In France in 1998, a 45-year-old female was admitted to hospital after a suicide attempt involving a salad made of five handfuls of oleander leaves, alongside two different types of sleeping tablets.[35] She was discharged a few days after admission, but only after all the oleander leaves had been pumped from her stomach. In Italy in 2019 a 58-year-old female was found dead grasping an oleander branch. High concentrations of oleandrin were found in her blood, but the main evidence was the oleander plant material that was found on her tongue and in her stomach contents.[73]

Accidents with whole oleander plants also happen. In Italy in the early 2000s, an apparently homeless couple was found dead in the woods. Oleander leaves were found near the bodies and in their stomachs. Their identities remained a mystery for four years until they were recognised on Belgian television. They had chosen to live a vegan lifestyle (rare at that time) and it's thought they ate the plants by accident while foraging for food.[58]

It's not just humans who can fall foul of oleander. In spite of the leaves' bitter taste, animals are still tempted to nibble on these plants,[32] or can be poisoned if they have accidentally been added to their feed.[74,75] This is particularly a problem for grazing animals where there is a scarcity of edible plants. In fact, in India oleander was known as *Kajamaraka*, which translates as "the herb that makes the horse die".[13] Animal poisoning is much more difficult to diagnose than human poisoning – apart from the obvious communication barrier, the symptoms tend to be very non-specific and the chemical culprit is usually only identified in the stomach contents of poisoned animals, once it's too late, or sometimes spotted in the feed by a sharped-eyed farmer.

Oleander poisoning can be passed up the food chain to humans. In 2006 in Italy there was an unusual case of a 43-year-old female and a 66-year-old male falling ill after eating snails from their garden that had fed on the leaves of their oleander plants. Both were suffering from gastrointestinal and cardiac symptoms when they arrived at the hospital, but later made a full recovery.[76] As well as the human patient's body fluid samples, some leftover frozen snails were also analysed – this use of insects in toxicology is a relatively new field called "forensic entomotoxicology."[74]

### 7.7.2   Datura

Deliberate suicide attempts with Datura have also been seen, with 30 being recorded by the Swiss poisons centre between 1995 and 2009, although this number was small compared to the 290 taking Datura for recreational reasons.[72] Between 2002 and 2017, 6 suicide attempts with Datura were reported to the Austrian poisons centre.[77]

Datura is a tempting recreational drug because of the hallucinations users can get after drinking Datura tea or smoking its leaves.[36] This means many of the toxicology cases reported involve younger patients.[4] As the staff of the Poisons Information Centre in Austria discovered between 2002 and 2017 – the most likely group to be trying plants for pleasure were youths aged 15 to 20 years, with 20 cases of Datura poisoning reported to them. Interestingly the most popular plant for abuse was its cousin, Angel's trumpet (*Brugamansia suaveolens*), which we met in Chapter 2.[67] One hospital in the USA had 10 cases of Datura psychosis over a three-month period, including two 15-year-old boys found by police naked and delirious wandering around a field after eating five or six flowers.[78]

In Australia in 1999 five teenage males (aged 14 to 15 years), mixed boiled Datura leaves with Coca-Cola then drank it on the school bus. Within an hour they were in A&E with dilated pupils, acting aggressively and plucking at imaginary objects. After 36 h they were released with only three of them requiring medical interventions such as sedatives or even a ventilator.[79]

During the summer of 2006, four teenagers (aged 13 to 16) were hospitalised in Canada after eating Datura seeds. All four were hallucinating and so aggressive they had to be restrained by staff, but recovered after two days.[80]

In a somewhat alarming case from Germany in 2006, an 18-year-old drank a cup of tea made by boiling two flowers of Angel's Trumpet. A few hours later he was so psychotic that he amputated his own penis and tongue with a pair of pruning shears. Although he was taken to hospital, unfortunately the detached parts couldn't be re-implanted. Fortunately (for him, if not the witnesses) he awoke after surgery with no memory at all of the events.[36] Incidents like these that are fatal would likely be recorded by a Coroner as "misadventure" and are sometimes the outcome of teenage pranks gone wrong. Unlike many recreational drugs, Datura is particularly tempting to adolescents as they view it as a "free drug growing in the park" or in neighbouring gardens.[80]

There are also cases where foraging children chew on the seeds accidentally. In a case in Armenia in 2020, a 5-year-old child was taken to hospital after eating a plant and starting to hallucinate and behave oddly. The plant was found and identified by a botanist as Datura, and after three days she was discharged from hospital fully recovered.[81] Whereas adults would likely be sedated with **diazepam**, treating agitated children is more tricky, and drugs are not recommended straight away.[4] As any parent of a toddler will know, some young children will also just spit out any oral medications, so need to be treated with an injection instead.[80]

Sometimes accidental poisonings happen when Datura plants are inadvertently harvested with crops and ground into flour.[8] In 2019, 315 adults became ill and another five adults died in Uganda after eating cereal provided by the World Food Programme that had been contaminated with Datura.[82] There was another recent example of this in France where Datura-contaminated buckwheat flour was behind several calls to the local poison centre.[83]

As we saw in Chapter 2, scopolamine has been associated with drug-facilitated crimes such as robbery or sexual assault. Passengers on public transport may fall into the trap of accepting food, drink, or tobacco from apparently friendly strangers that actually contains Datura extract.[4]

### 7.8   CATCHING THE CHEMICAL CULPRIT

The low concentrations of oleandrin that we are looking for in forensic toxicology cases (0.001 mg L$^{-1}$) mean that we often need very sensitive methods.[84] **High-pressure liquid chromatography (HPLC)** can be used to detect oleandrin and its breakdown product oleandrigenin,[4] and it was used in the case of a one-year-old boy taken to hospital in Canada in 1997. He was suffering from vomiting, was lethargic and (unsurprisingly) was crying inconsolably. Fortunately he recovered after two days but it remained a mystery where he got the plant from.[85]

We saw in Chapter 2 how the technique **LC-MS** works, and the first part (the separation process) is the same in HPLC. As each liquid component of the mixture comes out of the machine, it passes into another one called a **diode array detector** (DAD), which tries to identify it. The DAD works by shining light through the liquid, some of which is absorbed and some of which passes through. The specific kind of light used is ultraviolet light (the same kind that is so damaging in sunlight) and this is useful as many naturally occurring chemicals absorb this type of light, but over very slightly different ranges, allowing us to tell them apart. As we don't always know the exact number to choose in advance, the DAD scans over a broad range, giving us the best chance of picking up each alkaloid. HPLC-DAD is not as selective as LC-MS as there can be several chemicals that absorb at the exact same point, making them hard to tell apart.

As oleandrin is a cardiac glycoside like the chemicals found in digitalis (see Chapter 6), there is a temptation for emergency doctors or clinical toxicologists to try to use **immunoassay** methods used to detect digoxin, when faced with an oleander poisoning. Some of these tests will give a positive result, as oleandrin partially cross-reacts with the antibodies for digoxin. The assay gives an apparent digoxin concentration, even if there is no digoxin in the samples.[86] This only gives an indication though that something similar to digoxin has been taken, and it is impossible to relate the result (an approximate number) to the symptoms.[35] Partial cross-reactivity usually means the concentration is an underestimate,[58] so the patient's symptoms should drive the treatment approach until a more specific test, such as HPLC can be done.

There is one way to tell if it is really digoxin, and that is to test the patient's potassium levels. As we saw in the previous chapter, high potassium levels are a consequence of digitalis poisoning, but they are not high after oleander overdose.[85] As the antibodies to digoxin can be 'fooled' into attaching themselves to oleandrin this means the life-saving antidote DigiFab® can also be used to treat oleander poisoning.[10,87]

## 7.9   CASE CLOSED

Towards the end of *White Oleander*, Astrid recalls the deadly draught her mother Ingrid brewed up as containing oleander, jimsonweed and belladonna (see Chapter 4). But in the case history at the start of this chapter Ingrid adds some heart-shaped flowers that are never identified (belladonna's flowers are not heart-shaped, see Figure 4.2)

She soaked a plant collected in moonlight from the neighbors' fence, with little heart-shaped flowers.

p. 36
*White Oleander* by Janet Fitch,[1] 1999

The most common heart-shaped flowers are known as "bleeding hearts" (*Dicentra spectabilis* or *Lamprocapnos spectabilis*), Figure 7.5. And although they are not well known as being poisonous, they do contain a small amount of the alkaloid **protopine**, a skin irritant[8,10] and are a poetic addition to Ingrid's revenge potion.

Most of the poisonings with oleander and Datura we have seen in this chapter were caused by eating, drinking or smoking the plants. Our case study however, was a "contact poisoning", where the poison was absorbed directly through the skin and then passed into the bloodstream, and this explains why the DMSO was needed. Our skin is very good at protecting us from potential poisons, particularly water-soluble ones,[88] but DMSO helps to carry drugs across the skin's barrier. We also know it dissolves the alkaloids in oleander as it is used by medical researchers looking at their effect on cells.[89] It has some unfortunate side-effects though including rash, nausea, garlic breath, headache and loss of taste.[19] As we don't know the timeline of Barry's demise, he may not have had time to notice these effects before falling seriously ill and dying. Skin reactions such as rashes tend

**Figure 7.5**    Bleeding hearts (*Lamprocapnos spectabilis*).

to show more quickly than ill-effects after eating a poison (particularly if absorbed through the thicker skin of the hands), but we would expect Barry's symptoms to appear within a few hours at most.[58] As we saw earlier, Datura has a history of being applied topically in Ayurvedic medicine.

If his death was not witnessed and Barry was found dead some time later, it then would be down to the crime scene investigators to collect the key evidence. Unfortunately, unless there are very obvious clues pointing towards drugs or poisons, such as the remains of a fire (Chapter 12), empty pill blister packets, used needles or bottles labelled with a skull and crossbones, things can be missed. For teenagers who die as a result of inhaling solvents (a "sudden sniffing death") for example, cans of deodorant or air freshener in their bedrooms are often ignored despite being the cause of death, as they don't look out of place. If contact poisoning was not suspected in Barry's case straight away, it's unlikely the investigators would have swabbed the doorknobs or furniture looking for traces of plant poisons.

Ingrid has spent six years in prison for murder, by the time we learn a little more about Barry's death. She contacts some legal campaigners and hopes to appeal on the grounds that the proper procedures weren't followed at her trial

She was denied due process. It's in the record...
The public defender didn't even raise a sweat in her defense....
The prosecution's case was completely circumstantial...
Barry Kolker could have died of heart failure,...
The autopsy was not conclusive. He was overweight, and a drug user,

p. 342–343
*White Oleander* by Janet Fitch,[1] 1999

This tells us that there was no confirmation that plant poisons were involved – possibly that post-mortem toxicology was not done at all in the case. And although the autopsy was inconclusive (not uncommon in poisonings) it suggests that his heart was affected in some way, if Ingrid's new defence lawyers could pass off his death as a heart attack. As we have seen in our real cases, the heart is often affected by oleander and Datura, and death comes more easily to those already not in the best of health.

## REFERENCES

1. J. Fitch, *White Oleander*, Virago, London, 2000.
2. J. A. Vale, T. C. Marrs and R. L. Maynard, *Clin. Toxicol.*, 2018, **56**, 1093–1097.
3. P. White, *Crime Scene to Court: The Essentials of Forensic Science*, RSC Publishing, Cambridge, 4th edn, 2016.
4. V. V. Pillay and A. Sasidharan, *Indian J. Crit. Care Med.*, 2019, **23**, S250–S255.
5. R. J. Slaughter, M. G. Beasley, B. S. Lambie, G. T. Wilkins and L. J. Schep, *N. Z. Med. J.*, 2012, **125**, 87–118.
6. P. Dioscorides, *The Greek Herbal of Dioscorides*, Haner Publishing, New York, 1959.
7. Medical Economics Company, *PDR for Herbal Medicines*, Medical Economics Company, Montvale, NJ, 1998.
8. M. R. Cooper, A. W. Johnson and E. A. Dauncey, *Poisonous Plants and Fungi: An Illustrated Guide*, The Stationery Office, Norwich, 2nd edn, 2003.
9. V. Bandara, S. A. Weinstein, J. White and M. Eddleston, *Toxicon*, 2010, **56**, 273–281.
10. D. Frohne and H. J. Pfänder, *Poisonous Plants: A Handbook for Doctors, Pharmacists, Toxicologists, Biologists and Veterinarians*, Manson Publishing, London, 2nd edn, 2004.
11. M. Blamey and C. Grey-Wilson, *Mediterranean Wild Flowers*, Harper Collins, London, 1993.
12. U. S. Praveen, M. D. Gowtham, C. V. Yogaraje-Gowda, V. G. Nayak and M. B. Mohan, *Int. J. Med. Toxicol. Forensic Med.*, 2012, **2**, 135–142.
13. L. Ceci, F. Girolami, M. T. Capucchio, E. Colombino, C. Nebbia, F. Gosetti, E. Marengo, F. Iarussi and G. Carelli, *Toxins*, 2020, **12**, 471.
14. Y. Gaillard and G. Pepin, *J. Chromatogr. B: Biomed. Sci. Appl.*, 1999, **733**, 181–229.
15. P. Dey, *Biomed. Pharmacother.*, 2020, **129**, 110422.
16. A. Carfora, R. Petrella, R. Borriello, L. Aventaggiato, R. Gagliano-Candela and C. P. Campobasso, *Forensic Sci., Med., Pathol.*, 2021, **17**, 120–125.
17. S. P. Nordt, M. Hendrickson, K. Won, M. J. Miller, S. P. Swadron and F. L. Cantrell, *Am. J. Emerg. Med.*, 2020, **38**, 1698. e1695–1698.e1696.

18. C. M. Skinner, *Myths and Legends of Flowers, Trees, Fruits, and Plants: In All Ages and in All Climes*, J. B. Lippincott, Philadephia, 1911.
19. *Disposition of Toxic Drugs and Chemicals in Man*, ed. R. C. Baselt, Biomedical Publications, Seal Beach, 12th edn, 2020.
20. L. M. Blum and F. Rieders, *J. Anal. Toxicol.*, 1987, **11**, 219–221.
21. W. Turner, *A New Herball*, Cambridge University Press, Cambridge, 1989.
22. J. Gerard and T. Johnson, *The Herball or Generall Historie of Plantes*, Adam Islip, Joice Norton & Richard Whitakers, London, 2nd edn, 1636.
23. W. Wojtyna and F. Enseleit, *Pacing Clin. Electrophysiol.*, 2004, **27**, 1686–1688.
24. A. Foster, *The Medicinal Plant Collection at the University of Oxford Botanic Garden*, Wellcome Trust, Oxford, 2010.
25. A. Chevallier, *Encyclopedia of Medicinal Plants*, DK Publishing, St Leonards, 2001.
26. L. S. Nelson, M. A. Howland, N. A. Lewin, S. W. Smith, L. R. Goldfrank and R. S. Hoffman, *Goldfrank's Toxicologic Emergencies*, McGraw Hill, New York, 11th edn, 2018.
27. B. Hubbard, *Poison: The History of Potions, Powders and Murderous Practitioners*, Welbeck, London, 2020.
28. K. Fatur, *Econ. Bot.*, 2020, **20**, 1–19.
29. R. Bevan-Jones, *Poisonous Plants: A Cultural and Social History*, Windgather Press, Oxford, 2009.
30. D. Frohne and H. J. Pfänder, *A Colour Atlas of Poisonous Plants*, Wolfe Publishing Ltd, London, 1983.
31. M. Brown, *Death in the Garden: Poisonous Plants & their Use Throughout History*, Pen & Sword Books Ltd, Barnsley, 2018.
32. A. Stewart, *Wicked Plants*, Algonquin Books of Chapel Hill, Chapel Hill, NC, 2009.
33. M. Grieve, *A Modern Herbal*, 3rd edn, Tiger Books International, Twickenham, 1998.
34. *Gerard's Herbal*, ed. M. Woodward, Senate, London, 1994.
35. A. Tracqui, P. Kintz, F. Branche and B. Ludes, *Int. J. Leg. Med.*, 1997, **111**, 32–34.
36. A. Marneros, P. Gutmann and F. Uhlmann, *Eur. Arch. Psychiatry Clin. Neurosci.*, 2006, **256**, 458–459.
37. J. L. Müller, *J. Toxicol., Clin. Toxicol.*, 1998, **36**, 617–627.

38. M. Chamberlin, *Old Wives' Tales: The History of Remedies, Charms and Spells*, The History Press, Cheltenham, 2020.
39. D. A. Driggers, R. Solbrig, J. F. Steiner, J. Swedberg and G. S. Jewell, *West. J. Med.*, 1989, **151**, 660–662.
40. J. P. André, *J. Chem. Educ.*, 2013, **90**, 352–357.
41. *A New Grove Dictionary of Opera*, ed. S. Sadie, Oxford University Press, Oxford, 1997.
42. C. E. Burden, *Libretto Lakme: An Opera in Three Acts*, Steinway, New York, 1900.
43. S. T. Dietz, *The Complete Language of Flowers: A Definitive and Illustrated History*, Wellfleet Press, New York, 2020.
44. C. Meldrum, *Madapple*, Alfred A Knopf Inc, New York, 2008.
45. J. Suchard and A. Greb, *J. Med. Toxicol.*, 2021, **17**, 57–60.
46. A. Seton, *Dragonwyck*, Mariner Books, Boston, USA, 1944.
47. Toxic Trees: Folklore of Juniper, Laburnum and Oleander, https://www.icysedgwick.com/toxic-trees/, accessed February 2022.
48. S. Schultes, A. Hofmann and C. Rätsch, *Plants of the Gods*, Healing Arts Press, Rochester, VT, 2001.
49. A. J. Carter, *Br. Med. J.*, 1996, **313**, 1630–1632.
50. F. Inkwright, *Botanical Curses and Poisons*, Liminal 11, London, 2021.
51. M. Kaplan, in *Discover Magazine*, Kalmbach Media, Waukesha, 2015.
52. T. Stone and G. Darlington, *Pills, Potions, Poisons: How Drugs Work*, Oxford University Press, Oxford, 2000.
53. F. Inkwright, *Folk Magic and Healing: An Unusual History of Everyday Plants*, Liminal 11, London, 2019.
54. H. V. Harissis, *Perspect. Biol. Med.*, 2014, **57**, 351–360.
55. Anonymous, *The Poison Garden*, The Alnwick Garden, Alnwick, 2005.
56. A. Dey and J. N. De, *Afr. J. Tradit., Complementary Altern. Med.*, 2011, **9**, 153–174.
57. Z. Küçükdurmaz, H. Karapınar, I. Gül and A. Yılmaz, *Arch. Turk. Soc. Cardiol.*, 2012, **40**, 168–170.
58. L. Papi, A. B. Luciani, D. Forni and M. Giusiani, *Am. J. Forensic Med. Pathol.*, 2012, **33**, 93–97.
59. S. K. Manna, N. K. Sah, R. A. Newman, A. Cisneros and B. B. Aggarwal, *Cancer Res.*, 2000, **60**, 3838–3847.
60. S. Pathak, A. S. Multani, S. Narayan, V. Kumar and R. A. Newman, *Anti-Cancer Drugs*, 2000, **11**, 455–463.

61. M. Hermel, M. Sweeney, Y.-M. Ni, R. Bonakdar, D. Triffon, C. Suhar, S. Mehta, S. Dalhoumi and J. Gray, *J. Evidence-Based Integr. Med.*, 2021, **26**, 1–16.

62. Ryn, in *Botany After Dark*, Podtail, Stockholm, 2021.

63. L. Ma, R. Gu, L. Tang, Z.-E. Chen, R. Di and C. Long, *Toxins*, 2015, **7**, 138–155.

64. J. Zhou, G. Xie and X. Yan, *Encyclopedia of Traditional Chinese Medicines*, Springer, Heidelberg, 2011.

65. Banned and restricted herbal ingredients, https://www.gov.uk/government/publications/list-of-banned-or-restricted-herbal-ingredients-for-medicinal-use/banned-and-restricted-herbal-ingredients, accessed May 2020.

66. A. Kerchner and Á. Farkas, *Forensic Toxicol.*, 2020, **38**, 30–41.

67. S. Dorner-Schulmeister, K. Bartecka-Mino and A. Holzer, *40th International Congress of the European Association of Poisons Centres and Clinical Toxicologists (EAPCCT)*, Tallinn, Estonia, 2020, p. 536.

68. M. Un, C.-K. Chen, L. Labat and B. Mégarbane, *40th International Congress of the European Association of Poisons Centres and Clinical Toxicologists (EAPCCT)*, Tallinn, Estonia, 2020, p. 546.

69. J. Pietsch, R. Oertel, S. Trautmann, K. Schulz, B. Kopp and J. Dreßler, *Int. J. Leg. Med.*, 2005, **119**, 236–240.

70. I. A. Wasfi, O. Zorob, N. A. Al katheeri and A. M. Al Awadhi, *Forensic Sci. Int.*, 2008, **179**, e31–e36.

71. B. E. Haynes, H. A. Bessen and W. D. Wightman, *Ann. Emerg. Med.*, 1985, **14**, 350–353.

72. J. Fuchs, C. Rauber-Lüthy, H. Kupferschmidt, J. Kupper, G.-A. Kullak-Ublick and A. Ceschi, *Clin. Toxicol.*, 2011, **49**, 671–680.

73. E. Azzalini, M. Bernini, S. Vezzoli, A. Antonietti and A. Verzeletti, *J. Forensic Leg. Med.*, 2019, **65**, 133–136.

74. R. Chophi, S. Sharma, S. Sharma and R. Singh, *J. Forensic Leg. Med.*, 2019, **67**, 28–36.

75. E. R. Tor, D. M. Holstege and F. D. Galey, *J. Agric. Food Chem.*, 1996, **44**, 2716–2719.

76. C. Gechtman, F. Guidugli, A. Marocchi, A. Masarin and F. Zoppi, *J. Anal. Toxicol.*, 2006, **30**, 683–686.

77. S. Dorner-Schulmeister, K. Bartecka-Mino and A. Holzer, *40th International Congress of the European Association of Poisons Centres and Clinical Toxicologists (EAPCCT)*, Tallinn, Estonia, 2020, p. 537.

78. R. C. W. Hall, M. K. Popkin and L. E. McHenry, *Am. J. Psychiatry*, 1977, **134**, 312–314.
79. P. D. Francis and C. F. Clarke, *J. Paediatr. Child Health*, 1999, **35**, 93–95.
80. T. H. Wiebe, E. S. Sigurdson and L. Y. Katz, *Paediatr. Child Health*, 2008, **13**, 193–196.
81. B. Mkhitaryan, G. Mazmanyana, H. Ghazaryan, M. Grigoryan, H. Apresyan, V. Asoyan and M. Grigoryan, *40th International Congress of the European Association of Poisons Centres and Clinical Toxicologists (EAPCCT)*, Tallinn, Estonia, 2020, p. 545.
82. W. A. Abia, H. Montgomery, A. P. Nugent and C. T. Elliott, *Compr. Rev. Food Sci. Food Saf.*, 2021, **20**, 501–525.
83. M. Glaizal, C. Schmitt, L. Tichadou, J.-M. Sapori, M. Hayek-Lanthois and L. de Haro, *Presse Med.*, 2013, **42**, 1412–1415.
84. F. Gosetti, C. Nebbia, L. Ceci, G. Carelli and E. Marengo, *Anal. Methods*, 2019, **11**, 5562–5567.
85. A. Gupta, P. Joshi, S. A. Jortani, R. Valdes Jr, T. Thorkelsson, Z. Verjee and S. Shemie, *Ther. Drug Monit.*, 1997, **19**, 711–714.
86. A. Dasgupta, S. A. Risin, M. Reyes and J. K. Actor, *Am. J. Clin. Pathol.*, 2008, **129**, 548–553.
87. R. M. Janssen, M. Berg and D. H. Ovakim, *Can. Med. Assoc. J.*, 2016, **188**, 747–750.
88. K. Harkup, in *The Guardian*, Guardian Media Group, London, 2016.
89. N. Amend, F. Worek, H. Thiermann and T. Wille, *Toxicol. Lett.*, 2021, **350**, 261–266.

# CHAPTER 8

# The Power to Heal as Well as Harm

If you see a term that's **bold** it's defined in the Glossary. Only the first time that the word appears in the chapter will it be indicated in this way.

---

**Case History: Misdiagnosis**

*A rich elderly lady is taken ill with a fever, not long after the sudden death of her husband. The next morning her symptoms worsen and she dies suddenly and painfully despite treatment from a doctor.*

| | |
|---|---|
| DOCTOR: | Did you notice the symptoms of the disease to which Madame de Saint-Méran has fallen a victim? |
| VILLEFORT: | I did. Madame de Saint-Méran had three successive attacks, at intervals of some minutes, each one more serious than the former. When you arrived, Madame de Saint-Méran had already been panting for breath some |

*(continued)*

---

Poisonous Tales: A Forensic Examination of Poisons in Fiction
By Hilary Hamnett
© Hilary Hamnett 2023
Published by the Royal Society of Chemistry, www.rsc.org

minutes; she then had a fit, which I took to be simply a
nervous attack, and it was only when I saw her raise herself
in the bed, and her limbs and neck appear stiffened, that
I became really alarmed. Then I understood from your
countenance there was more to fear than I had thought.
This crisis past, I endeavored to catch your eye, but could
not. You held her hand — you were feeling her pulse —
and the second fit came on before you had turned towards
me. This was more terrible than the first; the same nervous movements were repeated, and the mouth contracted
and turned purple. And at the third she expired. At the end
of the first attack I discovered symptoms of tetanus; you
confirmed my opinion.

p. 1618–1619

*The Count of Monte Cristo* by Alexandre Dumas,[1] 1844

## 8.1   THE INVESTIGATION

Our case history describes the death of an old lady at home in
bed. In modern times, someone this unwell would be taken to
hospital for treatment, but in the 1800s, when this book is set,
hospitals were a place to be avoided by the wealthy. Partly because
early hospitals had poor people in them, but also because they
were for convalescing and recovery and not for dying or infec-
tious people.[2] It was therefore much more common for people
who could afford it to be treated at home by their family doctor,
as we see in Madame Saint-Méran's case. For those who could not
afford a doctor, they may have sought the help of a neighbour-
hood 'old wife', the female custodians of community domestic
medicine.[3]

At first, the witnesses (the doctor and the victim's son-in-law
Villefort) presume the death has been caused by tetanus, but
later the doctor reveals he suspects it was poisoning by "vegetable substances" possibly **brucine** or **strychnine**. The death of
Mr Saint-Méran, her husband, a short time earlier had already
been explained away as an apoplectic stroke, but his wife was in
much better health. She also 'imagines' seeing someone enter
her room in the dead of night and hears them touching her
glass of orangeade. The doctor knows there are deadly poisons
in the household as he is using them to treat another resident

(Mr Noirtier), and worries that a servant has accidentally poisoned Madame Saint-Méran

| | |
|---|---|
| DOCTOR: | May not Barrois, the old servant, have made a mistake, and have given Madame de Saint-Méran a dose prepared for his master? |
| VILLEFORT: | ...Yes. But how could a dose prepared for M. Noirtier poison Madame de Saint-Meran? |
| DOCTOR: | Nothing is more simple. You know poisons become remedies in certain diseases, of which paralysis is one. For instance, having tried every other remedy to restore movement and speech to M. Noirtier, I resolved to try one last means, and for three months I have been giving him brucine; |

p. 1622
*The Count of Monte Cristo* by Alexandre Dumas,[1] 1844

However, Barrois himself is the next victim of the poisoner after accidentally drinking Mr Saint-Méran's remaining [bitter-tasting] poisoned lemonade and suffering a similar set of symptoms

Barrois, his features convulsed, his eyes suffused with blood, and his head thrown back, was lying at full length, beating the floor with his hands, while his legs had become so stiff, that they looked as if they would break rather than bend. A slight appearance of foam was visible around the mouth, and he breathed painfully, and with extreme difficulty.

p. 1788
*The Count of Monte Cristo* by Alexandre Dumas,[1] 1844

There are two more non-fatal poisonings in the book, one with strychnine or brucine, and one with an un-named narcotic, and we will explore these later.

So who is behind these poisonings? The perpetrator is the second wife of Villefort – who is trying to secure the fortune of the Saint-Mérans for herself and her son. But the idea of poisoning was actually planted by the protagonist of the story, the Count. Running through the complex plot of *The Count of Monte Cristo* is the theme of revenge. The protagonist is 19-year-old Edmond Dantès, who is falsely imprisoned without trial for treason and sent to a remote French island fortress. While there, he discovers who has set him up and escapes to the treasure island of Monte Cristo where he finds enough wealth to buy himself a new identity as a Count. He seeks revenge on his captors including the corrupt judge who convicted him, Villefort. An opportunity comes when Villefort's daughter Valentine and her [evil] stepmother battle over the inheritance. The Count is behind the plot, having given the idea of poisoning with brucine to the stepmother.

## 8.2  THE PLANT BEHIND THE STORY

The two poisons suspected in the triple poisoning are brucine and strychnine, both of which are found in the *Strychnos nux-vomica*, *Strychnos tiente* (Upas) or *Strychnos ignatii* trees.[4] They are native to Southeast Asia, Australia and Hawaii and are also known as "poison nut", "Quaker buttons", "semen strychnos", and the "vomiting nut".[5] *Strychnos nux-vomica* is a tall evergreen tree with glossy oval leaves, tubular, unpleasant-smelling white flowers and when ripe yellow/orange fruit about 5 cm in diameter[6] (Figure 8.1). The fruit contains a white jelly-like pulp and exceptionally bitter disc-shaped seeds covered in hair.[5,7] The seeds are the most poisonous parts of the plant containing 3% alkaloids, with strychnine and the less toxic brucine the most concentrated and found in roughly equal amounts.[6,8] Strychnine is also found in the bark and the dried blossom.[9] *Strychnos toxifera*, is one of the plant sources of the poison curare, which was used to

**Figure 8.1**    A Strychnos nux-vomica plant. © fritzifoto/123RF.COM.

poison arrow heads for hunting, another being *Chrondrodendron tomentosum.*[10]

For many people, strychnine is what comes to mind when they think of poisonings.[11] This may be to do with its long history as an arrow poison,[12] or the several high-profile true crimes that involved it in the 19th Century (the heyday of strychnine). But, it could also be its dramatic role in popular murder mysteries by writers such as Agatha Christie[13,14] and Arthur Conan Doyle.[15]

The effects of strychnine have been well-known since the times of Ancient China and India.[9] The plants were brought to Europe in the 15th Century initially as a rodent poison in Germany, but weren't used in medicine until 1640. In the *Count of Monte Cristo*, we see the doctor trying to use brucine to cure Mr Noirtier's paralysis. By the 1800s, strychnine and brucine were isolated from the plant[11] and were being made up into 'tonics' in small amounts to treat muscle tone and improve circulation.[16] While strychnine was popular in the UK, brucine was more common in France, where our murders are set.[17] They were also sold in over-the-counter as laxatives and for colic in babies,[18] and as 'pep up' tonics for the elderly. The idea stemmed from its bitter taste – this stimulates saliva and increases appetite in a similar way to the bitter **quinine** in tonic water.[9] In Australia in the late 1800s, snake-bite envenomation was treated with injections of strychnine.[19] Medical use of strychnine continued up to the 1970s, when doctors finally agreed that it had no health benefits.[12,20]

Strychnine itself is not very soluble in water,[21] so it is usually reacted with other chemicals to make it into a more soluble form called a "salt". As we have already learned, alkaloids are alkaline, so reacting the plant with acids neutralises them forming a salt that can be crystallised out as a solid.[22] The three most common salts were sulphate, nitrate and hydrochloride.[23]

When collected for medicinal use (mainly homeopathy now), the discs are dried in the sun and powdered. Once in this form they can unfortunately be easily mistaken for other powders, such as olive or date nuts.[6]

After consuming strychnine there is no fixed time interval before effects start, but it is usually 10–15 min.[11] As we see with Madame Saint-Méran, many victims are apparently well then suddenly fall ill. The effects are initially tremors and twitching, staggering when walking followed by life-threatening spasms. These are caused by strychnine taking over control of the nervous system – flicking a nerve 'on switch' called the **glycine** receptor.[10] Tremors and spasms are caused by strychnine over-exciting nerves in the body, causing muscles to contract. The characteristic strychnine spasms happen when the powerful muscles at the back of the body contract, causing the spinal cord to extend. The body then rests on the back of the head and on the heels (which can sometimes touch in extreme cases when the body arches up[11]). These convulsions last for 1–2 min and are followed by a quiet period of 10–15 min before coming on again. Profuse sweating and thirst also occur and the patient remains conscious throughout (strychnine stimulates the brain giving it a heightened sense of perception[24]), in intense pain. Light and noise can bring on the convulsions, so moving the victim to a quiet, dark room can help.[11] Symptoms of poisoning can occur after ingestion of just one bean.[6]

The cause of death is usually asphyxia due to paralysis of the muscles around the chest and diaphragm. Death usually occurs within 1–3 h and can occur during a spasm or between them, meaning the patient may be found contorted in the pose of their final convulsion, or in a relaxed position.

Another effect of strychnine is to lower the pH of the blood (making it more acidic). This happens because **lactate** levels rise during all that muscular activity.[25] It is not the only chemical to do this – and this lowering of pH, which is called "acidosis" is also seen by forensic toxicologists in deaths due to chronic (long-term) alcoholism and poorly controlled diabetes.

There is no antidote to strychnine,[26] so the only way to treat strychnine poisoning is to stop the convulsions before they exhaust the muscles and paralyse the victim's breathing. **Diazepam** is the most effective drug for this,[27] and though it is better known for calming anxiety, it is also prescribed for muscle problems such as sciatica. In extreme cases, the patient may need a general anaesthetic. It is then a matter of waiting for the poison to pass through the body.[12]

The name "vomiting nut" suggests strychnine might cause an upset stomach, but as we have seen, vomiting is not one of the main symptoms of strychnine poisoning. Turner writing in the 1500s suggests mixing the seeds with water, salt, honey, and dill leaves and drinking the concoction to bring on vomiting.[28] This might seem a curious idea, but by Turner's day "purging" was a well-established, and often ineffective, treatment for all sorts of ailments.[2] The idea of quickly removing poisons from the body is a good one, still in use today. For example, if a patient has only just swallowed the strychnine, doctors may try gastric lavage (pumping the stomach) to remove any unabsorbed poison. But purging only works if the poison has been swallowed; we will see in Chapter 9 that traditional remedies for snakebite venoms involving vomiting and diarrhoea do not work, and can lead to dangerous delays in seeking treatments that do.

In the *Count of Monte Cristo*, the idea of gradually building up **tolerance** (see Chapter 2) to brucine with small doses is suggested by the Count to the poisoner, Madame de Villefort

"Well," replied Monte Cristo "suppose, then, that this poison was brucine, and you were to take a milligramme the first day, two milligrams the second day and so on...Well, then, at the end of a month, when drinking water from the same carafe, you would kill the person who drank with you, without your perceiving, other than from slight inconvenience, that there was any poisonous substance mingled with this water.

p. 1166
*The Count of Monte Cristo* by Alexandre Dumas,[1] 1844

In the end, this strategy is used against the poisoner, when she tries to murder her stepdaughter, Valentine. Although she leaves a large dose of poison in Valentine's glass, the girl survives because she has been advised to take a smaller dose of brucine every day to prepare her for the attempt on her life. But actually the opposite is true of the alkaloids in *Strychnos* trees. Instead of building up immunity or resistance, something called **sensitisation** happens. Every time strychnine is taken, even in small amounts, the body becomes more sensitive to it, meaning the symptoms are actually stronger rather than weaker.[29] It used to be thought that the drug was simply building up in the body until a fatal dose was reached,[6] but we now know that in non-fatal cases, strychnine is broken down (metabolised) by the body and excreted in the urine.[16] Strychnine can build up in people with liver damage though, as most drug metabolism happens in the liver.[6]

### 8.3   POISONOUS PLOTS

Probably the most iconic consequence of strychnine poisoning is the 'mask' or sardonic grin that arises from the face muscles going into spasm. This has given it the name "the smiling poison".[29] Commentators think that the made-up "strych-nodide", responsible for the Joker's grin in *Batman* is really strychnine.[30] Whilst the "rictus grin" is real, the idea that it's a long-term effect in the living is not; patients who survive strychnine poisoning are unlikely have symptoms after a few weeks.[31]

We see the fatal consequences of strychnine in what looks like a murder–suicide in the Alfred Hitchcock film *Psycho* (1960) set in 1959. The teenage Norman Bates' mother is found dead in bed with her lover and a suicide note

SHERIFF:      ...it's the only murder-and-suicide case in Fairvale led-
              gers! Mrs Bates poisoned this guy she was ... involved
              with, when she found out he was married, then took
              a helping of the same stuff herself. Strychnine. Ugly
              way to die.

p. 94
*Psycho* screenplay by Joseph Stefano,[32] 1959

It turns out that 15 years on, Norman has disinterred her remains, preserved her body using taxidermy and continues to hold 'conversations' with her. He was also responsible for the double poisoning (along with several other murders by this point in the plot), although we don't see any of the symptoms in the film.

A more recent film, Wes Anderson's *The Grand Budapest Hotel* (2014), also features a strychnine poisoning. Monsieur Gustave, the concierge of a mountainside resort is framed for the murder of a wealthy older woman with whom he's been having an affair. In this scene, the hotel's lawyer is reading the accusation from the woman's family

DEPUTY
KOVACS:      The next morning, Madame D. was found dead by
             strychnine poisoning. M. Gustave was not
             observed on the premises again until," of
             course, "twenty-four hours later." The
             identity of his accusers is made clear in
             this notarized deposition.

p. 47
*The Grand Budapest Hotel* best original screenplay
by Wes Anderson,[33] 2013

It's not clear how the authorities came to their cause of death, but an image of the deceased is printed in a newspaper and shown at the inquest. On screen she is seen lying dead on a tiled floor but not in a characteristic strychnine pose.

In the fantasy novel *The Anubis Gates* (1983), Professor Brendan Doyle an expert on poetry, gets stuck in 1801 after time travelling. He begins hallucinating after eating something

It smelled vaguely like curry, but seemed to be some kind of cold stew made of leaves and something that looked like kiwi fruit, but smaller and harder and more furry...
He'd seen them before, in the Foster Gardens of Nuuanu in Hawaii, on a tall tree whose scientific name he still remembered: *Strychnos Nux*
*Vomica*, the richest source of raw strychnine.
He'd been eating strychnine.

p. 228
*The Anubis Gates* by Tim Powers,[34] 1983

Doyle remembers a story about a man recovering from strychnine poisoning after consuming activated charcoal (see Chapter 3), and so eats a bowlful of ashes and incinerated wood from a fireplace. We have seen already that delirium and hallucinations are not symptoms of strychnine poisoning. But even if this were strychnine, charcoal straight from the fireplace, known as "raw" charcoal, is much less effective as a treatment for poisoning than its "activated" cousin. To turn raw charcoal into a medicine it has to go through an activation process, involving grinding the charcoal down into a fine dust and then treating it with steam, oxygen, carbon dioxide and acids. This removes impurities and gives it a huge surface area to bind to drugs or poisons, something clinical toxicologists call "decontamination".[35,36] It doesn't work on all poisons, it is useless against cyanide (Chapter 10) for example, and it only captures poison that is still in the digestive system – once a poison has passed into the blood it's too late for charcoal. It certainly does work against strychnine though, and the story Doyle remembers may well have been the hair-raising real public demonstration by a man called Pierre Touéry in front of the French Academy of Medicine in 1831.[35] Determined to impress, he swallowed 10-times the lethal dose of strychnine along with 15 grams of activated charcoal and survived.[36]

As well as being used in emergency medicine, activated charcoal is now also sold as a "detox powder" in health food shops, and even in vegan croissants that are claimed will detoxify your body.

The collection of poems called *A Shropshire Lad* by A. E. Housman published in 1896, mentions strychnine in a deadly but unsuccessful poisoning attempt on Mithridates (see Chapter 5)

They put arsenic in his meat
And stared aghast to watch him eat;
They poured strychnine in his cup
And shook to see him drink it up;
They shook, they stared as white's their shirt:
Them it was their poison hurt.
—I tell the tale that I heard told.
Mithridates, he died old.

p. 94
*A Shropshire Lad* by A. E. Housman,[37] 1896

In the sci-fi novel *The Invisible Man* by H. G. Wells (also published in the late 19th Century, like the Housman poem) we see strychnine's reputation as a 'tonic'. Griffin, a scientist who is trying to make himself invisible, has worked out the secret mixture of drugs that will do it, but needs to recover before he is ready to try the concoction on himself

After a time I crawled home, took some food and a strong dose of strychnine, and went to sleep in my clothes on my unmade bed. Strychnine is a grand tonic, Kemp, to take the flabbiness out of a man.

p. 53
*The Invisible Man* by H. G. Wells,[38] 1897

A somewhat extreme version of this tonic features in the collection of short stories, poems and essays *Shandygaff* (1918) by Christopher Morley. In the story, *The Haunting Beauty of Strychnine*, he describes a fictional French town called Strychnine where patients could take a "strychnine bath" on their doctor's orders

After two weeks of the strychnine baths the merry monarch [Edward VII] is said to have called for a corncob pipe and a plate of onions, after which he made his escape by walking over the forest track to the French frontier, although previous to this he had not walked a kilometer without a cane

p. 139
*Shandygaff* by Christopher Morley,[39] 1918

There is little evidence of these strychnine baths actually existing, and we know strychnine itself is not very soluble in water, but it can absorb through the skin, as we will see in some of the toxicology cases later in this chapter. The short story also doesn't say what the strychnine has cured, but given the description of the king's walking, possibly tuberculosis.

In Charles Dickens' novel *Martin Chuzzlewit* (1843) there are two poisonings, both connected to Martin's cousin Jonas. In the first, Jonas' wealthy father Anthony is suddenly taken ill

... looking round, they saw Anthony Chuzzlewit extended on the floor, with the old clerk upon his knees beside him...He had fallen from his chair in a fit, and lay there, battling for each gasp of breath, with every shrivelled vein and sinew starting in its place.
They raised him up, and fetched a surgeon with all haste, who bled the patient and applied some remedies; but the fits held him so long that it was past midnight when they got him—quiet now, but quite unconscious and exhausted—into bed.

Chapter 18
*The Life and Adventures of Martin Chuzzlewit* by Charles Dickens,[40] 1843

Anthony dies shortly after this, and we find out later in the book that Jonas had talked previously to a doctor about poisoning his father by adding something to his cough mixture. The same doctor later procures him two different "drugs", one of which is slow acting and not suspicious in appearance. Although we don't know what the drugs were, these symptoms (a period of convulsions followed by exhaustion and death) are very much consistent with strychnine poisoning. Jonas is later arrested for murder and eventually takes his own life with another poison (see Chapter 10).

An apothecary (see Chapter 5) supplies one of the three men in *The Pardoner's Tale* (1392) with the poison he uses to kill his drinking mates. The three have found a stash of treasure, but turn on each other in their greed. Two of them stab the youngest man then celebrate by drinking the wine he has brought them, without realising it is poisoned. The only description of the poisoning Chaucer gives us is that they "died on the spot", but earlier, the apothecary, believing he is selling rat poison to the man, describes the effects

A thing so strong that, as my soul's to save,
In the whole world there is no living creature
Which, if it swallows any of this mixture,
No bigger amount than a grain of wheat,
But must then lose its life upon the spot;
Yes, it must die, and that in a less while,
Believe me, than it takes to walk a mile;
This poison is so strong and virulent.

p. 338–339
*The Pardoner's Tale* by Geoffrey Chaucer,[41] 1392

Strychnine is a tempting choice for this poison, but as we have seen, it was not introduced to Europe until the 15th Century (about a century after this tale was written). It could have been one of the plants we met in earlier chapters, but none of them is quite so dramatic as to make someone die on the spot, so this was most likely a made-up rat poison.

Back in *The Count of Monte Cristo*, Valentine's second bout of being poisoned is reminiscent of Shakespeare's Juliet (see Chapter 4). Valentine's stepmother tries a new poison, realising that the old one wasn't going to work, replacing the drink on her bedside with a powerful narcotic. However, Valentine has already been given a non-fatal dose of a narcotic by the Count in the form of a pastille, and it is this that makes her fall into a deep sleep

The young girl no longer breathed, no breath issued through the half-closed teeth; the white lips no longer quivered — the eyes were suffused with a bluish vapor and the long black lashes rested on a cheek white as wax.

p. 2189
*The Count of Monte Cristo* by Alexandre Dumas,[1] 1844

The Count intercepts the stepmother's poison and throws away some to make it look like Valentine has drunk it and become her latest victim. Just like Juliet, a funeral is held for Valentine and she is taken to a cemetery in Paris (with customary yew trees, see Chapter 3) before being rescued by the Count. We do not know what was in the pastille that gave her such a convincing appearance of death, but it could have been either of the poisons we met in Chapter 2 (**morphine** or **scopolamine**) or even **heroin**; before it was recognised as being dangerous and addictive, heroin was sold as cough pastilles.

There is some interesting forensic chemistry at the scene of Valentine's death, when the family doctor tries to test the liquid in the glass by her bedside using a few drops of nitric acid. The liquid goes blood-red and he has a eureka moment, although without telling anyone else why. This was, in fact, an early presumptive test for opiates such as morphine, heroin and **codeine**. At crime scenes, investigators can test suspicious powders or liquids using *spot tests* or *colour tests*. A small volume of a chemical or a mixture of chemicals is dropped onto the sample, creating a colour. The right colour change tells them the sample needs further investigation in the lab, and it can even tell them the possible family of drugs present. These tests cannot be used in Court though as they can give false positives (like the **immunoassays** we saw in Chapter 6). They can also be fooled by cutting agents added to street drugs, and the colour change itself is a bit subjective. The doctor at Valentine's 'death' watches the liquid turn red when he adds the nitric acid, suggesting the presence of morphine (Figure 8.2).[42] It was not until 20 years after *The Count of Monte Cristo* was written, that a presumptive test for strychnine was published. It was a similar process but used peroxide and sulphuric acid to create the red colour.[43]

Strychnine's fame as a poison for animals appears in *All Creatures Great and Small* written by James Herriot in 1972. By this time, strychnine was no longer being used medicinally in the UK, but was a common rat and mole poison. One tactic was to soak earthworms in a strychnine solution and put them down mole holes.[44] All that changed in 2004 when the European Commission banned strychnine for use in pest control because of the suffering it causes to animals (they have the same symptoms as humans).[45] Once the UK left the European Union at the end

**Figure 8.2**   Establishing the presence of the poison, from The Count of Monte Cristo.[44]

of 2020, the rules around possession of strychnine in the UK reverted to the Poisons Act 1972, and those wishing to buy poisons including strychnine, arsenic, mercury (see Chapter 11), thallium and lead now require a licence from the Home Office.[46]

## 8.4   MODERN MEDICAL USES

In Traditional Chinese Medicine (TCM), a paste of Strychnos seeds is used externally to relieve pain, lower fevers, treat various types of tumor and to relieve paralysis.[6,7] Today, many prescriptions use a processed (heated) formulation of Nux vomica powder, for example, *Mi qian* zi, with a strychnine content of approximately 0.8%.[23,47] Strychnos seeds, known as *Tumdgha*, are commonly used by traditional Tibetan healers to treat traumatic injuries, pains, anaesthesia, paralysis, and tumors.[48] The processing usually involves boiling or frying the seeds, but doesn't always result in a safe remedy – sometimes alkaloids are converted into their nitrogen oxidation derivatives (*e.g.*, brucine-*N*-oxide), which can also be poisonous.[48] Deaths from TCM remedies containing Strychnos plants happen, particularly when prescribed by non-professional herbalists.[49]

Extracts of Strychnos fruits are often called "slang nuts" (in Cambodia) or "slang Chai nut" (in Thailand) and are used to treat gastrointestinal problems and even malaria. In a case in the USA in 2017, a 40-year-old Cambodian woman drank what she thought was vodka, but it turned out to be a herbal remedy containing approximately 1% strychnine that was being stored in a vodka bottle. She was taken to hospital with abdominal cramps but discharged a few hours later.[26] Slang nuts can also be wrapped in tobacco and leaves and held against the inside of the cheek, but in 1995 in the USA, a 58-year-old woman went to hospital with cramps and dizziness after chewing one instead. She recovered after five days.[50]

A surprisingly common homeopathic remedy, the Strychnos plant (known as "Nux vom") is used for digestive problems,[4] sensitivity to cold in 'chilly individuals' and irritability.[5,7] Some homeopathic practitioners recommend it for children who have eaten too much rich food at a birthday party![51] Typical homeopathic remedies do not contain any measurable amount of active ingredient, they are made by diluting a "mother tincture", which can itself be toxic. In a case in France 2012, a 22-year-old woman swallowed 200 mL of a Nux vom mother tincture and was taken to hospital with muscle spasms, but went on to recover within a couple of days.[52]

### 8.5  MODERN TOXICOLOGY CASES

Fortunately, because of legal controls,[20] the characteristic symptoms of strychnine poisoning, and its easy chemical detection, it is no longer the poison of choice.[53] Although in Sri Lanka in 2011, a 35-year-old woman died within a few hours of arriving at hospital with convulsions. Her husband was arrested and admitted to hiding it in empty amoxicillin capsules. He died by injecting himself with strychnine powder before he could be sentenced.[54]

Interestingly, strychnine is so stable that it can still be detected in human remains from decades earlier. In a case in Romania in 2018, the remains of a 25-year-old man who had been buried in 1975 were exhumed after his family became suspicious. The initial cause of death had been ruled as suffocation, but after the remains were analysed, strychnine and delorazepam (a similar drug to diazepam) were found.[55]

Forensic toxicology on human remains is quite unusual, but exhumations can be ordered by the Courts, requested by family members, or done at the behest of insurance companies.[56] This is usually when new evidence has come to light, as we saw in the Harold Shipman case in the UK in 2000, when some of his victims were exhumed and tested for **morphine** (see Chapter 2). It's not just suspected poisonings or intoxications that can trigger exhumations, sometimes it may be cases of medical negligence or when it may finally be possible to identify a Jane or John Doe. Usually by the time a body is exhumed, all of the body fluids have dried up, and the toxicologist has to make do with tissue, bones or hair. While these can tell us if a drug or poison is present, any levels measured can't be easily related to blood levels. In the case of strychnine this doesn't really matter, as just being present in the body is suspicious.[57] Other drugs can be detected in partially decomposed bodies, such as arsenic, **digitoxin** (Chapter 6), mercury (Chapter 11), and thallium.[56]

An altogether trickier situation is when toxicology is needed on an embalmed body. During embalming, body fluids are replaced by formaldehyde mixed with water and methanol. Formaldehyde is well-known for reacting with nitrogen-containing chemicals, which as we have seen includes the alkaloids found in poisonous plants. These reactions can either decrease or increase drug concentrations, meaning they are no longer a reliable measure of what was in the person's system at the time of death. In some cases, the drug reacts with the formaldehyde forming metabolites (breakdown products) giving the impression the drug was taken much earlier than it was, or at a lower dose. In other cases, existing metabolites can be turned back into the drug, giving the opposite impression.[58] This can cause problems for exhumed bodies, but also for those not yet buried but who have been embalmed very quickly after death – a common custom in hot countries.

Before the substance was banned, poisoning by strychnine was mostly seen by forensic toxicologists as a suicide method. In 2002 in the UK, a 42-year-old man went to hospital after drinking a bottle of wine along with some white powder from his shed. Within a few minutes of arriving, he developed a marked tremor and muscular spasms and shortly after this had a respiratory and cardiac arrest so was taken to ICU. After three days he was still twitching, but went on to recover completely.[25] Of course, even after the ban, old bottles of strychnine remained in garden sheds and in rural areas, so a few cases were still seen.[59] In France, authorities tried to improve the safety of pest control products containing strychnine by insisting the dye methylene blue was added to them (we also see this tactic with antifreeze). This distinctive colour alerted investigators to the suicide of a 60-year-old man in 2015, after he was found dead with a glass of blue liquid near him. Strychnine was confirmed in his post-mortem blood sample at a concentration of 25 mg L$^{-1}$, over 100 times the concentration at which other fatalities have been reported (>0.2 mg L$^{-1}$ [60]).

In North America, strychnine can still be bought as "gopher bait" in low concentrations (0.3–3%)[61] and most of their poisoning cases come from exposure to baits, including through the skin.[62] In 2004 in the USA, a 52-year-old man was found dead after drinking a liquid containing gopher pellets.[61] Occasionally there are deaths due to strychnine-contaminated heroin,[63] although given they are usually in long-term heroin users, these may be missed and recorded as simple heroin overdoses. Strychnine is added to heroin as a bulking agent because it mimics the bitter taste.[64] Something that catches toxicologists out occasionally is when a tourist brings in and takes something that is banned or just not prescribed in the country they are visiting. This happened in the mysterious case of the Body on the Moor[65] in the UK in 2015, when a man took strychnine he had brought in from Pakistan, where it is still used legally for pest control.

In South America, poisoning with pesticides is still a leading case of hospital admissions. In one case in 1999 in Chile, a 6-year-old boy was brought to A&E unconscious and suffering contractions after drinking a solution of strychnine his father had made to kill wild dogs. He was put into a coma to stop the contractions and made a full recovery.[66]

The stimulant effect of strychnine was thought to help athletes, and so it was used as an early form of sports doping,[67] and remains on the World Anti-Doping Agency's list of banned substances even in 2022.[68] In a case in 2018 in the USA, a 15-year-old teenager developed muscle spasms and needed CPR after drinking a protein shake that had been deliberately contaminated with strychnine (the motive remains a mystery).[69]

The generic appearance of strychnine as a white powder means it can be mistaken for recreational drugs. In Ireland in 1982, 8 young men snorted strychnine believing it to be cocaine. Within 30 min they had been taken to hospital and one died of a cardiac arrest.[70]

Although the symptoms of strychnine poisoning are so iconic, misdiagnosis can still happen even in the 21st Century. In an interesting case in the USA in 2020, a 19-year-old male went home from hospital, after being admitted for alcohol intoxication. He was given goldenseal (*Hydrastis canadensis*, Figure 8.3) tea to drink by his parents, probably because one of its reported health benefits is cleansing or treating liver problems.[7,71] A few hours later, he started to experience muscle contractions so strong that they caused him to arch his back off the bed, giving the appearance of strychnine poisoning. Goldenseal contains the alkaloid **hydrastine**, which chemically looks nothing like strychnine, but can also produce muscle spasms.[72]

**Figure 8.3**    A flowering goldenseal (*Hydrastis canadensis*) plant.

## 8.6  CATCHING THE CHEMICAL CULPRIT

Strychnine, like all the alkaloids we have seen in this book, contains nitrogen and so can easily be detected using a method called **gas chromatography** with **nitrogen–phosphorus detection (GC–NPD)**.[25] This combines the gas chromatography we saw in Chapter 3, with a kind of detector that only 'sees' nitrogen and phosphorous atoms. It works by burning the drug or poison in an air–hydrogen flame to produce charged particles known as "ions". The electricity produced by these ions is then measured by the detector as an electric current. The GC–NPD was a favourite of many forensic toxicology labs for years, as it cuts out a lot of the 'noise' in the results, such as compounds in the blood like cholesterol that don't contain nitrogen, but can hide other things. However, there are some important drugs that these detectors miss, such as the date-rape drug GHB, which doesn't contain nitrogen or phosphorous.

Because strychnine is much less available now, it may be missed during an investigation, particularly if the case is not associated with rural occupations like vets, farmers, and game-keepers.[73] The powerful muscle contractions it causes can also give the impression that a violent struggle has taken place. This happened in the case of a 46-year-old famer's wife in Scotland who was later found to be positive for strychnine.[74]

## 8.7  CASE CLOSED

The two fatal strychnine or brucine poisonings in *The Count of Monte Cristo* have very realistic symptoms. However, the extremely bitter taste of strychnine makes it a poor choice for poisoning by mouth. Both victims (Madame de Saint-Méran and Barrios the servant) take in the poison in a soft drink, making it implausible that they wouldn't have noticed the addition. A better choice would have been something already very bitter, like tonic water.[75]

Is it realistic that the first two deaths are mistaken in the novel for natural causes or disease? Unless you're naturally suspicious, or a forensic toxicologist, you don't necessarily assume that every death has involved drugs or poisons. In fact, the Office for National Statistics in the UK publishes causes of death data on a monthly basis, and the most common causes of death are

Alzheimer's or dementia followed by Ischaemic heart disease,[76] so a natural death is much more likely. In the 1800s, diseases and microbes were still not well understood, and there was no vaccine for tetanus at the time, so it was much more common.[77] It was also a good guess by the witnesses, as tetanus produces a toxin called "tetanospasmin", which can cause similar symptoms to those Madame Saint-Méran suffered: lockjaw; spasms of the face, neck and back; paralysis; and eventually death.[78] In some real-life strychnine poisonings, the symptoms have been initially mistaken for epilepsy or cholera, before murder was suspected.[29]

## REFERENCES

1. A. Dumas, *The Count of Monte Cristo*, The Floating Press, Waiheke Island, 2009.
2. *The Cambridge History of Medicine*, ed. R. Porter, Cambridge University Press, Cambridge, 2006.
3. M. Chamberlin, *Old Wives' Tales: The History of Remedies, Charms and Spells*, The History Press, Cheltenham, 2020.
4. D. Frohne and H. J. Pfänder, *Poisonous Plants: A Handbook for Doctors, Pharmacists, Toxicologists, Biologists and Veterinarians*, Manson Publishing, London, 2nd edn, 2004.
5. M. Castro, *The Complete Homeopathy Handbook: A Guide to Everyday Health Care*, Macmillan, London, 1990.
6. Medical Economics Company, *PDR for Herbal Medicines*, Medical Economics Company, Montvale, NJ, 1998.
7. A. Chevallier, *Encyclopedia of Medicinal Plants*, DK Publishing, St Leonards, 2001.
8. S. Funayama and G. A. Cordell, *Alkaloids: A Treasury of Poisons and Medicines*, Elsevier, Amsterdam, 2015.
9. J. Patocka, in *Handbook of Toxicology of Chemical Warfare Agents*, ed. R. C. Gupta, Academic Press, Boston, 2nd edn, 2015, ch. 17, pp. 215–222.
10. A. Stewart, *Wicked Plants*, Algonquin Books of Chapel Hill, Chapel Hill, NC, 2009.
11. J. Timbrell, *The Poison Paradox: Chemicals as Friends and Foes*, Oxford University Press, Oxford, 2005.
12. R. Bevan-Jones, *Poisonous Plants: A Cultural and Social History*, Windgather Press, Oxford, 2009.
13. K. Harkup, *A is for Arsenic: The Poisons of Agatha Christie*, Bloomsbury, London, 2015.

14. C. Valentine, *Murder Isn't Easy: The Forensics of Agatha Christie*, Sphere, London, 2021.
15. J. F. O'Brien, *The Scientific Sherlock Holmes: Cracking the Case with Science & Forensics*, Oxford University Press, New York, 2013.
16. *Disposition of Toxic Drugs and Chemicals in Man*, ed. R. C. Baselt, Biomedical Publications, Seal Beach, 12th edn, 2020.
17. C. J. S. Thompson, *Poison Romance and Poison Mysteries*, George Routledge & Sons, Ltd, London, 1904.
18. Anonymous, *The Poison Garden*, The Alnwick Garden, Alnwick, 2005.
19. A. Mueller, *On Snake-Poison: its Action and its Antidote*, L. Bruck Medical Publisher, Sydney, 1893.
20. J. Emsley, *More Molecules of Murder*, RSC Publishing, Cambridge, 2017.
21. *Clarke's Analysis of Drugs and Poisons*, ed. A. C. Moffatt, D. Osselton and B. Widdop, Pharmaceutical Press, 4th edn, London, 2011.
22. T. Hargreaves, *Poisons and Poisonings: Death by Stealth*, RSC Publishing, Cambridge, 2017.
23. R. Guo, T. Wang, G. Zhou, M. Xu, X. Yu, X. Zhang, F. Sui, C. Li, L. Tang and Z. Wang, *Am. J. Chin. Med.*, 2018, **46**, 1–23.
24. B. Hubbard, *Poison: The History of Potions, Powders and Murderous Practitioners*, Welbeck, London, 2020.
25. D. M. Wood, E. Webster, D. Martinez, P. I. Dargan and A. L. Jones, *Crit. Care*, 2002, **6**, 456–459.
26. T. Singhapricha and A. C. Pomerleau, *J. Emerg. Med.*, 2017, **52**, 493–495.
27. T. Stone and G. Darlington, *Pills, Potions, Poisons: How Drugs Work*, Oxford University Press, Oxford, 2000.
28. W. Turner, *A New Herball*, Cambridge University Press, Cambridge, 1989.
29. F. Inkwright, *Botanical Curses and Poisons*, Liminal 11, London, 2021.
30. D. A.-S. Jürgens, P. D. Tscharke and P. J. Brocks, *J. Graph. Nov. Comics*, 2022, **13**, 685–699.
31. I. Makarovsky, G. Markel, A. Hoffman, O. Schein, T. Brosh-Nissimov, Z. Tashma, T. Dushnitsky and A. Eisenkraft, *Isr. Med. Assoc. J.*, 2008, **10**, 142–145.
32. J. Stefano, *Psycho Screenplay*, 1959.

33. W. Anderson, *The Grand Budapest Hotel Best Original Screenplay*, Twentieth Century Fox, Los Angeles, CA, 2014.
34. T. Powers, *The Anubis Gates*, Gollancz, London, 1983.
35. R. W. Derlet and T. E. Albertson, *West J. Med.*, 1986, **145**, 493–496.
36. R. Ashton and C. Leblanc, *Dalhous. Med. J.*, 2010, **37**, 29–31.
37. A. E. Housman, in *A Shropshire Lad*, George, G. Harrap & Co. Ltd, London, 1969, ch. LXII, pp. 91–94.
38. H. G. Wells, *The Invisible Man: A Grotesque Romance*, Jazzybee Verlag, Loschberg, 2017.
39. C. Morley, *Shandygaff: A Number of Most Agreeable Inquirendoes Upon Life and Letters, Interspersed with Short Stories and Skits, the Whole Most Diverting to the Reader*, Garden City Publishing Company, New York, 1918.
40. C. Dickens, *The Life and Adventures of Martin Chuzzlewit*, Chapman & Hall, London, 1843.
41. G. Chaucer, *The Canterury Tales*, Oxford University Press, Oxford, 1998.
42. A. Choodum and N. Nic Daeid, *Talanta*, 2011, **86**, 284–292.
43. D. Lindo, *Assoc. Med. J.*, 1856, **s3–4**, 966.
44. M. Brown, *Death in the Garden: Poisonous Plants & their Use throughout History*, Pen & Sword Books Ltd, Barnsley, 2018.
45. European Commission, ed. E. Union, EC, Brussels, 2004, vol. 91/414/EEC.
46. Licensing for home users of poisons and explosive precursors, https://www.gov.uk/government/publications/licensing-for-home-users-of-explosives-precursors/licensing-for-home-users-of-poisons-and-explosive-precursors, accessed March 2022.
47. J. Otter and J. L. D'Orazio, *Strychnine Toxicity*, StatPearls Publishing LLC, Treasure Island, FL, 2021.
48. L. Ma, R. Gu, L. Tang, Z.-E. Chen, R. Di and C. Long, *Toxins*, 2015, **7**, 138–155.
49. Y. G. Zhang and G. Z. Huang, *Am. J. Forensic Med. Pathol.*, 1988, **9**, 313–319.
50. J. Katz, K. Prescott and A. D. Woolf, *Am. J. Emerg. Med.*, 1996, **14**, 475–477.
51. Homoeopathic Development Foundation, *Homoeopathy for the Family*, Wigmore Publications Limited, London, 1988.

52. T. Gicquel, S. Lepage, A. Baert and I. Morel, *Ann. Toxicol. Anal.*, 2012, **24**, 33–37.

53. J. H. Bock and D. O. Norris, in *Forensic Plant Science*, ed. J. H. Bock and D. O. Norris, Academic Press, San Diego, 2016, ch. 1, pp. 1–22.

54. S. Kodikara, *J. Forensic Leg. Med.*, 2012, **19**, 40–41.

55. G. P. Bonete, C. P. Martínez, F. L. Lucia and A. Luna, *Rom. J. Leg. Med.*, 2018, **26**, 298–301.

56. W. Grellner and F. Glenewinkel, *Forensic Sci. Int.*, 1997, **90**, 139–159.

57. R. J. Dinis-Oliveira, F. Carvalho, J. A. Duarte, F. Remião, A. Marques, A. Santos and T. Magalhães, *Toxicol. Mech. Meth.*, 2010, **20**, 363–414.

58. P. Nikolaou, I. Papoutsis, A. Dona, C. Spiliopoulou and S. Athanaselis, *Forensic Sci. Int.*, 2013, **233**, 312–319.

59. C. Duverneuil, G. L. d. l. Grandmaison, P. d. Mazancourt and J.-C. Alvarez, *Forensic Sci. Int.*, 2004, **141**, 17–21.

60. M. Schulz, A. Schmoldt, H. Andresen-Streichert and S. Iwersen-Bergmann, *Crit. Care*, 2020, **24**, 195.

61. T. Lindsey, J. O'Hara, R. Irvine and S. Kerrigan, *J. Anal. Toxicol.*, 2004, **28**, 135–137.

62. R. Greene and R. Meatherall, *J. Anal. Toxicol.*, 2001, **25**, 344–347.

63. M. Levine, A.-M. Ruha, K. Graeme, D. E. Brooks, J. Canning and S. C. Curry, *Chest*, 2011, **140**, 1357–1370.

64. L. S. Nelson, M. A. Howland, N. A. Lewin, S. W. Smith, L. R. Goldfrank and R. S. Hoffman, *Goldfrank's Toxicologic Emergencies*, McGraw Hill, New York, 11th edn, 2018.

65. J. Manel, in *Episode 3: Poison*, BBC, London, 2016.

66. B. Oberpaur, A. Donoso, C. Clavería, C. Valverde and M. Azócar, *Pediatr. Emerg. Care*, 1999, **15**, 264–265.

67. P. Van Eenoo, K. Deventer, K. Roels and F. T. Delbeke, *Forensic Sci. Int.*, 2006, **164**, 159–163.

68. The Prohibited List, https://www.wada-ama.org/en/prohibited-list, accessed February 2022.

69. M. Moss and S. Kusin, *American College of Medical Toxicology 2018 Annual Scientific Meeting*, Washington, DC, 2018, pp. 40–41.

70. W. G. O'Callaghan, N. Joyce, H. E. Counihan, M. Ward, P. Lavelle and E. O'Brien, *Br. Med. J.*, 1982, **285**, 478.

71. S. T. Dietz, *The Complete Language of Flowers: A Definitive and Illustrated History*, Wellfleet Press, New York, 2020.
72. J. A. Thompson, N. Mckeown and R. G. Henrickson, *American College of Medical Toxicology 2020 Annual Scientific Meeting*, New York, NY, 2020, p. 165.
73. J. M. Heiser, M. R. Daya, A. R. Magnussen, R. L. Norton, D. A. Spyker, D. W. Allen and W. Krasselt, *J. Toxicol. Clin. Toxicol.*, 1992, **30**, 269–283.
74. J. S. Oliver, H. Smith and A. A. Watson, *Med. Sci. Law*, 1979, **19**, 134–137.
75. J. H. Trestrail III, *Criminal Poisoning: Investigational Guide for Law Enforcement, Toxicologists, Forensic Scientists and Attorneys*, Humana Press, Totowa, 2000.
76. Monthly mortality analysis, England and Wales: December 2021, https://www.ons.gov.uk/peoplepopulationandcommunity/birthsdeathsandmarriages/deaths/bulletins/monthlymortalityanalysisenglandandwales/latest, accessed February 2022.
77. J. Postgate, *Microbes and Man*, Cambridge University Press, Cambridge, 4th edn, 2000.
78. M. T. Madigan, K. S. Bender, D. H. Buckley, W. M. Sattley and D. A. Stahl, *Brock Biology of Microorganisms*, Pearson, New York, 15th edn, 2018.

# Beautiful but Deadly

If you see a term that's **bold** it's defined in the Glossary. Only the first time that the word appears in the chapter will it be indicated in this way.

---

**Case History: Suicide**

*Egypt is occupied by the Romans, but the General in charge, Antony, has been neglecting his duties as he is too busy drinking and partying with the famously beautiful Queen of Egypt, Cleopatra. His actions lead to a rift with the other rulers of the Roman Empire and eventually to a battle, which he and Cleopatra lose. She fakes her own death, leading Antony to fall on his sword. But when she sees him die, Cleopatra resolves to kill herself with the venom of a snake rather than be captured as a prisoner of war.*

CLEOPATRA: This proves me base:
      If she first meet the curled Antony,
      He'll make demand of her, and spend that kiss
      Which is my heaven to have. Come, thou mortal wretch,
      *[To an asp, which she applies to her breast]*
      With thy sharp teeth this knot intrinsicate
      Of life at once untie: poor venomous fool

*(continued)*

---

Poisonous Tales: A Forensic Examination of Poisons in Fiction
By Hilary Hamnett
© Hilary Hamnett 2023
Published by the Royal Society of Chemistry, www.rsc.org

> CLEOPATRA:
> Be angry, and dispatch. O, couldst thou speak,
> That I might hear thee call great Caesar ass
> Unpolicied!...
> As sweet as balm, as soft as air, as gentle.
> O Antony!—Nay, I will take thee too.
> *[Applying another asp to her arm]*
> What should I stay—
> *[Dies]*
>
> Act V, scene II
>
> *The Tragedy of Antony & Cleopatra* by William Shakespeare,[1] 1606–7

## 9.1   THE INVESTIGATION

When the bodies of Cleopatra (Figure 9.1) and her ladies-in-waiting (Charmian and Iras) are found by the triumphant new leader of the Roman Empire, Octavius Caesar, there is no obvious cause of death

> OCTAVIUS:
> Bravest at the last,
> She levell'd at our purposes, and, being royal,
> Took her own way. The manner of their deaths?
> I do not see them bleed.
>
> Act V, scene II
> *The Tragedy of Antony & Cleopatra* by William Shakespeare,[1] 1606–7

All Octavius knows about the history is that a "simple country-man" was seen entering the place with what he said was a basket of figs. Their first thought is that these must have been poisoned (there are some fig leaves scattered around). But they discount this when they note that none of the women show any signs of swelling

> OCTAVIUS:
> O noble weakness!
> If they had swallow'd poison, 'twould appear
> By external swelling: but she looks like sleep
>
> Act V, scene II
> *The Tragedy of Antony & Cleopatra* by William Shakespeare,[1] 1606–7

It was thought, incorrectly, in Shakespeare's day that swelling was always seen in poisonings, but eventually someone spots the

**Figure 9.1**Death of Cleopatra, by Louis-Marie Baader.

puncture wounds on Cleopatra's breast and arm and realises it was a venomous snake.

So, what is the difference between a venom and a poison? A "venom" is a kind of poison that is secreted by an animal or insect with something pointy (fangs, stings, thorns or bristles).[2] These are known as "venomous animals". They can also spit venom into the eyes of their prey or aggressors.[3] There are also "poisonous animals", such as toads, but they need to be touched or ingested to cause symptoms (if you swallowed a venom it would likely be broken down by your stomach acid[4]). Some poisonous animals are brightly coloured to warn potential predators of their dangers,[5] whereas venomous animals like snakes are more adept at camouflage. Venoms are often used to attack prey, whereas poisonous plants and animals contain toxins and alkaloids as a defence mechanism.

How toxic a venom is, depends on the species involved and how it is delivered to its victim. A sting on the foot from a jellyfish may be less dangerous than a bite from a snake into a vein.[6] And even if you don't live in a far-flung and exotic location, you may well have experienced venom from an ant bite or wasp sting. Snake venoms differ from plant alkaloids as they are made up of much larger molecules – proteins. These work against the central nervous system or other tissues,[7] causing several different types of symptoms. There can be 2–200 different components in snake venom,[8,9] of which around 90% are proteins, with the rest being things like carbohydrates, lipids, metals (such as zinc[10]), amino acids and **enzymes** (to break down the flesh of the victim, making it easier to digest[4]).[11] The aim of the snake is not always to produce a rapid, dramatic death like Cleopatra's; some snake venoms are designed to kill quickly, whereas others slowly incapacitate the victim. We know of around 4000 species of snake currently in the World, of which 90% are thought to be venomous. But only around 200 of those are thought to be medically important, *i.e.*, able to cause harm to humans if bitten.[9,12]

In Cleopatra's case, the snake bite envenoming (SBE) was a suicide, but even in instances where it's an accident, SBE is often still fatal. This is partly because snakes are found in parts of the World where hospitals are few and far between, but also because victims may first go to a traditional healer or try a folk remedy.[13]

Even if an actual post-mortem examination is performed (rather than the cursory glance in Cleopatra's case) the Pathologist may not find very much. Autopsy findings in SBE are often non-specific and are consistent with a rapid death, which could have several different causes.[13] A firm cause of death in SBE relies on capture of the offending snake, finding toxins in biofluids (although as we will see, this is tricky), and the presence of fang marks on the body. The trouble is, if the area around the bite is severely swollen, these can be hard to see.[14] If bite marks *are* clear, the distance between the two fangs (usually a few cm, see Figure 9.2) can help identify the species of snake.[14] Getting the right species is especially important if the victim is still alive; as we will see later, there are antidotes to SBE but they are very specific, so choosing the correct one is key to survival.

## 9.2   THE SNAKE BEHIND THE STORY

All we know of Cleopatra's snake is that Shakespeare calls it an "asp" or a "worm". The most common species of snake in Egypt is the cobra (*Naja haje*, Figure 9.3) known as the "Egyptian Cobra" or "Brown Cobra" in English, but also by local names such as *Thaaban* and *Nachir*. It can found in Egypt, North Africa and also in many other countries in sub-Saharan Africa, preferably near some water.[15] *Naja haje* is a non-spitting cobra, and is considered

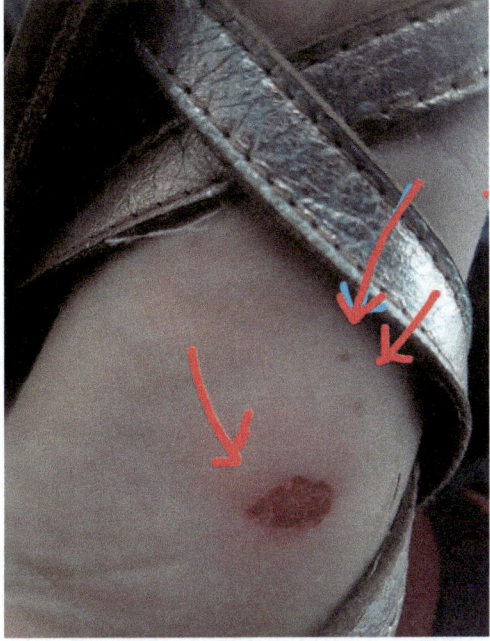

**Figure 9.2** A non-venomous snake bite on the foot received in Morocco, with arrows showing fang marks approximately 2 cm apart.

**Figure 9.3** An Egyptian Cobra (*Naja haje*).

one of the most deadly snakes in the World.[11] It can grow to 1–2 m in length, weigh 9 kg, and puts on an impressive defensive display by standing upright, hissing, and splaying out its 'hood', which can be 15–18 cm wide.[16,17] When not biting humans, *Naja haje* like to eat toads, small mammals, birds, eggs and smaller snakes.[15]

Snake venom is a modified type of saliva that is produced by special glands found in the upper jaw and can either run down a groove in the fangs or through an inner canal.[13]

When someone is bitten by a snake, a small reservoir of venom usually sits in the tissue under the skin and is taken up into the bloodstream gradually.[18] It typically contains a mix of toxic molecules that may attack cells (cytotoxins), the heart (cardiotoxins), blood (hemotoxins) and nerves (neurotoxins).[14,19] These toxins can act independently, together (known as "synergistically", see Chapter 3) or even against each other (where one toxin blocks another, known as "antagonistically"[20]).[21] Egyptian Cobras can hold around 350 mg of venom at any one time,[22] but it's difficult to predict how much venom has been passed to victim when they are bitten, as it depends on many factors. We will look at the different toxic molecules in more detail.

### 9.2.1 Cytotoxins

Cytotoxins punch holes in cell walls, causing them to die or even explode, resulting in pain and progressive swelling (known as "edema"), blistering at the bite site (see Figure 9.2), bruising and necrosis (rotting flesh that refuses to heal).[6,21] The breakdown in cell walls allows the venom to spread to other parts of the body.[23] The loss of so much fluid from the circulation into the bite area can then lead to shock.[4] *Naja haje* venom causes local edema, but not necrosis.[11]

### 9.2.2 Neurotoxins

Neurotoxins in the venom target the nervous system by preventing signals from being sent between nerves.[22] This can lead to intense pain everywhere in the body and not just at the bite, dizziness and convulsions[6,22] then paralysis. The paralysis usually starts in the eyelids then eventually spreads down the body to the chest, heart and lungs causing death by respiratory failure.[6,13,24]

Neurotoxicity can also make it hard for victims to speak or swallow and can lead to their vision becoming blurred.[17] These symptoms appear within 30 min of being bitten by an Egyptian Cobra, with death following within 2 to 16 h.[10]

### 9.2.3 Hemotoxins

Hemotoxins act on the cardiovascular system, producing coagulation disorders.[13] These disorders can either cause or prevent blood clotting. When blood can't clot properly, the victim bleeds profusely from the bite itself, but also internally. Some snake venoms dissolve the walls of blood vessels, causing bleeding under the skin, which appears as a bruise.[25] When the toxin causes blood to clot more than usual, it can lead to a stroke, quickly incapacitating the victim.

### 9.2.4 Cardiotoxins

As the name suggests, cardiotoxins interfere with the proper function of the heart. We met the same effect from some plant alkaloids in Chapter 6. The venom of the Egyptian Cobra affects the heart by blocking nerve signals being transmitted to it.[26]

As well as the symptoms described above, Egyptian cobra venom also contains myotoxins, which can lead to damage of blood vessels in the kidneys (nephropathy), and if untreated, acute (very quick and serious) renal failure.[21,27]

How serious SBE is, like all poisonings, depends on different factors such as where on the body the person was bitten (and at what angle), the size and species of the snake, the location, season, as well as the age, weight and health of both snake and victim.[13,21] The diet of the snake is also thought to play a role in the composition of its venom.[28] There is also a correlation between the length of the Egyptian Cobra and the dose of venom released, with longer snakes holding more venom.[22]

Once the person has been bitten, their body has little chance of fighting off the venom on their own using our natural immune system.[11] The good news is that there are specific antidotes to SBE called "antivenoms" or sometimes "antivenins" or "antisnake venoms".[7] These are commercially produced using other animals, such as horses or sheep, by injecting them with small (non-toxic) doses of venom over a long period.[11] The test animals produce antibodies to the venom in the same way as they would to a vaccination. They are then collected *via* a blood sample and isolated ready for use in humans.

Antivenoms are most useful when the exact species of the offending snake is known (because it's been photographed, filmed or captured at the scene for instance).[11] Symptoms can't be relied upon to identify the snake – they can only tell us what type of toxin was involved.[14] If the person was bitten by a native snake, there are some poly antivenoms (generic antidotes), which work against a mixture of toxins from the most common venomous snakes found in a certain region.[9,29] Antivenoms are given by intravenous injection,[30] and can be lifesavers, but do not reverse all the effects of SBE and some symptoms need to be treated with other drugs. They are also expensive, often need to be kept at low temperatures, and are not available in all parts of the World where there are venomous snakes. To add to this, some people will experience an allergic reaction to the antivenom. This makes these antidotes risky if accidentally used to treat a bite from a non-venomous snake,[31] or a "dry bite" from a venomous one (see Section 9.3).

Some folk remedies for SBE persist, often administered by traditional healers.[32] A few are very dangerous such as catching the snake and laying its head on the bite (likely to result in a second bite).[33] Others include applying a mixture of oil and warm milk to the bite mark, sucking out the venom by mouth or with a syringe, electric shock therapy, eating products that bring on vomiting and diarrhoea, and even tattooing, but sadly these do not work and can often cause new problems, or delays in lifesaving treatment.[15,30,33–35]

Even if antivenoms are available, there are other superstitions around catching and killing snakes, which can mean they are allowed to escape before they are identified, making it hard for doctors to pick the right antidote.[33] Unfortunately these folk remedies sometimes feature in popular fiction, helping them to live on (see Section 9.3).

Certain medicinal plants have been thought by traditional healers for centuries to have antitoxin properties. These include mistletoe,[33] *Strychnos nux-vomica* (Chapter 8) and *Nerium oleander* (Chapter 7),[36] as well as the chemicals **quercetin** (found in many vegetables) and **echinacoside** found in *Echinacea* plants

**Figure 9.4** (Left) Echinacea purpurea plants in flower, taken at Kew Gardens. (Right) The leaves of the plant with the link to treating snake bites on the sign, taken at the Chelsea Physic Garden. These plants are widely used in herbal medicine, particularly in the Americas for their effect on the immune system.[40,41]

(Figure 9.4).[32] In folk remedies for SBE, plant extracts are eaten, spread as a paste on the bite, or even mixed with urine and poured into the nostrils or ears of the victim.[37] However, although some of the compounds in these plants are promising, there are few studies that have rigorously tested their safety and efficacy. A recent exception was a cocktail of five plant extracts (*Azadirachata indica*, *Butea monosperma*, *Citrus limon*, *Clerodendrum serratum* and *Areca catechu*) in ethanol, which was evaluated as treatment for snake bites in India and found to be partially effective. The herbal cocktail worked against some of the snake venoms tested but not all, and the testing was carried out on cells and in animals rather than on a human SBE victim.[29]

## 9.3 VENOMOUS PLOTS

Our fascination with snakes, and the fear and suspense that the possibility of a bite brings, means they have been used in fiction for centuries. Snake stories can be found by Ancient Roman authors to the Bible and through to the modern day. Snakes are sometimes used to represent the sinister and evil, but also eternity, rebirth and renewal.

In Ovid's *Metamorphoses* book nine, the famously strong superhero Hercules kills Nessus the centaur with an arrow dipped in the venomous blood of the Hydra (a serpentine water monster he had killed in an earlier adventure).[38] The dying Nessus tells Hercules' wife Deïanira that centaur's blood, which has soaked into his tunic, can be used as a love potion (this is a lie and a cunning revenge plan by the centaur). When Hercules' interest in Deïanira begins to wane, she sends him the shirt as a gift without realising what the deadly consequences will be

The poisoned shirt was exposed to the heat, and its power was released
by the flames to creep on its cancerous way throughout Hercules' body...
He struggled at once to tear the lethal robe from his shoulders;
but where it yielded it tore at his skin...
Even his blood gave a hiss, like the sound of a plate of hot metal
plunged into icy water, and boiled in the fire of the poison.

Book 9, 160–170
*Metamorphoses* by Publius Ovidius Naso,[39] 8th Century CE

Hercules eventually dies an agonising death, but only after reeling off all of his triumphs in a loud voice first. It's not clear how much time has elapsed between the death of Nessus and the death of Hercules, but when venom is 'milked' from a snake (where they empty their glands in response to a squeeze of the head), it has to be immediately stored in a freezer at −20 °C to keep its biological activity, so it's not very plausible that without the help of magic, it would still be poisonous after drying onto a shirt.

Centuries later, venomous snakes remained a popular plot device. In Chapter 3 of this book, we met the three witches of *Macbeth* (1605) brewing up a magic potion that included hemlock. Before the poisonous plant is added, some bits of poisonous or venomous animals are thrown in

SECOND WITCH:     Eye of newt and toe of frog,
                  Wolf of bat and tongue of dog,
                  Adder's fork and blind worm's sting,
                  Lizard's leg and owlet's wing,

Act IV, scene i
*The Tragedy of Macbeth* by William Shakespeare,[40] 1605

Shakespeare seems to think that the venomous part of the adder is their tongue (as also mentioned in *Edward III* and a *Midsummer Night's Dream*),[6] but it is actually their retractable fangs.[41] He is aware of adder fangs or teeth however, as he mentions them in *Hamlet* (1600) and *Henry VI, Part 3* (1591–2). Assuming Shakespeare is talking about a native snake, it's likely to be the European adder (*Vipera berus*, Figure 9.5), the only venomous native snake to the UK, and found mainly in woodland. It bites an estimated 50–100 people every year, most of whom suffer only pain, bruising, and swelling. In a small number of cases, SBE symptoms such as low blood pressure, abdominal pain, diarrhoea, and collapse develop, and the victim needs to be treated with antivenom.[38] Of course there are also non-native snakes in

**Figure 9.5**   A European adder (*Vipera berus*) the only native venomous snake in the UK, taken at Jersey Zoo.

the UK kept as pets and in zoos and these bit 300 people between 2009 and 2020 (including 14 cobras).[42]

*The Jungle Book* (1894), a collection of short stories set in India by Rudyard Kipling, features several snakes. The most famous (thanks to Disney) is Kaa the Indian python (*Python molurus*) but as pythons are a non-venomous variety, there are no snake bites. In the short story called *Rikki-Tikki-Tavi*, a mongoose (a small mammal similar to a meerkat) kills a venomous common krait (*Bungarus caeruleus*) and outwits a pair of black cobras called Nag and Nagaina who are plotting to bite his human owners. Interestingly, the idea of folk remedies for SBE is also mentioned in this story

If you read the old books of natural history, you find they say that when the mongoose fights the snake and happens to get bitten, he runs off and eats some herb that cures him

p. 119
*The Jungle Book* by Rudyard Kipling,[43] 1894

Given the age-old link between snakes and sorcery, it's unsurprising that they make an appearance in five of the seven Harry Potter books: snake fangs are a potion ingredient in *The Philosopher's Stone* (1997);[44] a deadly basilisk bites Harry in *The Chamber of Secrets* (1998);[45] the pet snake of Lord Voldemort called Nagini is 'milked' in *The Goblet of Fire* (2000) and its venom mixed with unicorn blood to give him an almost-human form;[46] Mr Weasley is attacked by Nagini in *The Order of the Phoenix* (2003);[47] and both Harry and Professor Snape are bitten by the same snake in *The Deathly Hallows* (2007).[48]

When Harry is bitten on the arm by the basilisk (the mythical King of Serpents known since classical times[49]) he experiences some of the typical SBE symptoms we have seen above

The basilisk lunged again, and this time it's aim was true. Harry threw his whole weight behind the sword and drove it to the hilt into the roof of the serpent's mouth. But as warm [snake] blood drenched Harry's arms, he felt a searing pain just above his elbow...
White-hot pain was spreading slowly and steadily from the wound. Even as he dropped the fang and watched his own blood soaking his robes his vision when foggy. The chamber was dissolving in a whirl of dull colour.

p. 338
*Harry Potter and the Chamber of Secrets* by J. K. Rowling,[45] 1998

In the fantasy world of Harry Potter, he is saved by the tears of a Phoenix (a colourful magical bird[49]), which heal his wounds. Human tears have been found to have some therapeutic molecules, such as antibacterial proteins, but antivenom properties are still the stuff of magic.[50] The basilisk can also kill with its stare, something that Medusa, a female monster with hair made of venomous snakes in Greek mythology could also do.

In his second experience with a snake, Harry is bitten on the forearm by Nagini. Afterwards he falls unconscious and hallucinates, but the only lasting injury is two puncture wounds. Hermione cleans them up with some essence of dittany (*Dictamnus albus*) a plant considered in folk medicine to be an antidote to poison and a cure for venomous animal bites.[51] But none of the alkaloids in dittany, skimmianine, gamma-fagarine or **dictamnine** have been shown to have any antivenom properties, and the latter is actually toxic, causing dermatitis when the plant is touched.[51–53] Interestingly, the essential oils given off by dittany are known to ignite spontaneously in hot weather, giving it the folk name "burning bush" and making it a contender for the famous burning bush mentioned in the Bible.[54] So if dittany could not cure envenomation, how did Harry survive? If we suspend our disbelief for a moment, it's possible that this was a bite without venom – a "dry bite" – as approximately half of all snake bites are. These can happen because all the snake's stored venom was used for a previous bite, but can also be a deliberate choice by the snake to conserve venom (which takes a lot of energy to produce).[55]

Nagini definitely uses venom when she meets Mr Weasley. During the attack she strikes him three times on the chest, causing him to yell in pain, bleed and then pass out (likely due to loss of blood leading to shock). Mr Weasley is discovered and taken to hospital, where they try to treat his bitemarks

If only they could take the bandages off, I'd be fit to go home...
...I start bleeding like mad every time they try...
It seems there was some rather unusual kind of poison in that snake's fangs that keeps wounds open.

p. 451
*Harry Potter and the Order of the Phoenix* by J. K. Rowling,[47] 2003

We would expect profuse bleeding, as we have seen it is caused by the hemotoxins in real snake venoms interfering with blood clotting. Some snake bites in real life can also cause persistent wound re-infection, preventing them from healing, and requiring plastic surgery reconstruction.[56] The Healers (magical doctors) eventually discover an antidote to the venom and cure Mr Weasley, but we are not told what this is.

In the final and only fatal SBE in the Harry Potter series, Professor Snape is bitten by Nagini

Harry saw Snape's face lose the little colour it had left, it whitened as his black eyes widened, as the snake's fangs pierced his neck...
Snape, who fell sideways on to the floor, blood gushing from the wounds in his neck.

p. 536
*Harry Potter and the Deathly Hallows* by J. K. Rowling,[48] 2007

We are never told exactly what type of snake Nagini is, but at various points she is described as being 12 ft (about 3.5 m) long with the thickness of a man's thigh, and having a triangular head, a diamond-patterned tail and vertical slits for pupils. In the film adaptations she is shown as a Burmese or Reticulated Python, but all pythons are non-venomous, killing their prey by crushing or constricting it.

The second book of *A Series of Unfortunate Events* by Lemony Snicket (1999) is called *The Reptile Room* and features many snakes and a herpetologist. After the death of their parents in a house fire, the three Baudelaire children (Violet, Klaus and Sunny) are sent to live with a distant relative called Count Olaf. When it becomes clear he is trying to get hold of their inheritance, they escape to another relative, Dr Montgomery (Uncle Monty) the snake expert. But early one morning they find him dead in the room where he keeps his specimens

Klaus switched on one of the reading lamps to get a better look.
The shadowy mass was Uncle Monty. His mouth was slightly agape, as if he were surprised, and his eyes were wide open, but he didn't appear to see them.
His face, usually so rosy, was very, very pale, and under his left eye were two small holes, right in a line, the sort of mark made by the two fangs of a snake.

p. 265
*The Reptile Room* by Lemony Snicket,[57] 1999

A 'doctor' is called to do the autopsy (who arrives suspiciously quickly) and says he has found the venom of the fictional Mamba du Mal snake in Uncle Monty's veins, leading the adults to conclude that he has been accidentally bitten. Given that the autopsy took about 5 min, this seems a bit unlikely, and it turns out the 'doctor' is really one of the Count's henchmen. The real manner of death is unearthed by the children who discover that Uncle Monty has been manually injected with a sample of venom using a needle and syringe. Violet finds the evidence that the evil Count Olaf, who has been posing as Uncle Monty's new assistant, is behind the murder.

There are plenty of envenomations in the 2006 film *Snakes on a Plane*, including a non-fatal one of a small boy. The bite on his hand is seen by a flight attendant called Claire who uses a folk remedy to save the child

CLAIRE:     When I was a kid,
            Whenever we went hiking,
            We always carried olive oil and a razor blade in case of a
            snake bite.

*Snakes on a Plane* by David Dalessandro *et al.*[58] 2006

She coats the inside of her mouth with olive oil (to "seal it from the poison") then cuts into the bite area and sucks out the venom. There's little evidence of oil being able to act as a barrier like this, and we have already seen that snake venom is usually destroyed by stomach acid anyway. Cutting into a bite is definitely not a good idea as unless you're a highly skilled surgeon, you can end up cutting through a tendon, nerve or even artery.

## 9.4 SNAKES AND SPIRITUALITY

Serpents have been age-old symbols of magic, healing and evil.[59] In some cultures and religions snakes are worshipped and revered, whereas in others, they are regarded as manifestations of the Devil.

In rural parts of the USA, the handling of venomous snakes (typically rattlesnakes) is incorporated into some church services. Pastors in these churches believe that the Holy Ghost gives them the power to handle snakes and even drink deadly doses of **strychnine** (see Chapter 8) based on a Bible verse.[60] In Mark 16:

17–18 it says of Jesus' disciples: "they will pick up snakes in their hands, and if they drink any deadly thing, it will not hurt them."[61]

Another Bible verse suggests Christians will be protected against venomous snakes. In the Acts of the Apostles, Paul is bitten by a snake on the island of Malta

Paul had gathered a bundle of brushwood and was putting in on the fire, when a viper, driven out by the heat fastened itself on his hand...
He shook off the creature into the fire and suffered no harm.
They were expecting him to swell up or drop dead

Acts 28: 3–6
*The New Revised Standard* Bible[61]

Although vipers are definitely venomous, the identity of this particular snake remains a mystery, since there are no venomous snakes on Malta.[62] This could be a loose translation of the original Greek language of the Bible. The word ἔχιδνα could be translated as "viper" but also as "constrictor snake". Another theory is that the bite was from a Leopard snake (*Zamenis situla*) which is not actually venomous, with the reasoning that it *was* venomous when it bit him, but then there was a miraculous de-venoming of all snakes on Malta by Paul.[63] A more likely explanation is that it was a different island altogether and the name "Malta" has been mistranslated.

Despite various attempts to ban snake handling in churches over the years, it has carried on, and occasionally the death of a Pastor or member of the congregation is reported.[64]

In Africa and India, snake charming street performances can still be seen. Typically, they involve the charmer playing music to the snake to put it into a trance. In reality, although the snake appears to be moving in time to the music being played, it is actually just watching the swaying charmer.[65] Performances may also feature magic tricks, and dangerous acts like inserting a snake's head down the throat, eating a snake, or allowing one to bite the charmer to prove their immunity.[16] Snake charming has been associated with religious brotherhoods – and God's blessing is sometimes used to explain snake charmers' immunity. A more likely explanation is the charmers develop immunity over time after receiving several venomous but non-fatal bites (a concept we met with strychnine in Chapter 8),[4] or the snakes are milked before the show, making their bites less harmful for a while.[16] Other tactics can include using a non-venomous snake that looks to the untrained eye to be venomous, removing the snakes fangs or venom glands or even sewing the snake's mouth up.[66] Sometimes spectators are accidentally bitten during snake charming, including a 10-year-old, who was bitten on the hand when an Egyptian Cobra got out of control in Nigeria. He was given a herbal medicine, which made him vomit, but died about 2 h later.[30]

## 9.5   MODERN MEDICAL USES

Animal venom has been used for its therapeutic properties for thousands of years in countries such as China and Africa.[67] Western medicine was slow to catch on, only starting to investigate venoms in the 1870s, but by World War II several snake venom drugs had been approved for clinical use in the USA, for things like heart conditions, hemorrhages, pain, arthritis and cancer.[11] Some were abandoned because of safety concerns or lack of efficacy, but one of the most successful drugs developed was **captopril** from the venom of the Brazilian arrowhead viper (*Bothrops jararaca*, Figure 9.6). Victims of this viper's bite would often collapse as their blood pressure dropped – and this effect was harnessed to treat people suffering from high blood pressure (there are now lots of drugs on the market that do this, often

**Figure 9.6**    A Brazilian arrowhead viper (*Bothrops jararaca*).

ending in "-pril").[68] Captopril no longer comes directly from venom, but is made in a laboratory.

Snake venoms including from Egyptian Cobras are being investigated for treating cancers,[67] and controlling infections by working against microbes and parasites.[11,12,69,70]

The venom of snakes can also be used clinically to cause or prevent blood clots. Some venoms prevent the victim's blood from clotting and this can be used clinically, such as when treating or trying to prevent a stroke or heart attack.[71] Three drugs called eptifibatide from the pygmy rattlesnake (*Sistrurus miliarius*), Arvin® from the Malayan pit viper (*Calloselasma rhodostoma*), and tirofibran from the saw-scaled viper (*Echis* spp.) all prevent clotting. Sometimes the opposite effect is useful, for instance if a patient has haemophilia, they need help forming clots, and Russell's viper (*Daboia russelii*) venom causes very fast clotting (marketed as Stypven).[3] A more recent discovery in the USA was an enzyme, known as "reptilase" in the venom of the common Lancehea snake (*Bothrops atrox*). It can be made into a kind of biological glue to seal wounds and help blood to clot.[72]

## 9.6   MODERN TOXICOLOGY CASES

Venomous snake bites are often successfully treated in Western medicine in places such as Australia and the USA, but remain a neglected health problem in developing countries.[73] For forensic toxicologists, SBE may appear in accidents, homicides, suicides or even recreational use of venom. Although most cases are reported in India, working in a country without venomous snakes doesn't mean you won't encounter an SBE, as exotic snakes are kept as domestic pets and in zoos all over the World.[18]

Accidental snake bites occur when people deliberately handle snakes, but are also a common occupational hazard for farmers, plantation workers, and herders, with many non-fatal SBEs going unreported,[37] but leaving the victims with long-term physical injury such as amputation. In other cases, people happen upon snakes when walking (particularly at night) through fields and gardens. In China in 2020, a 20-year-old man was found in bed by his roommate with a swollen left thigh and dark purple, discoloured, blistered skin after walking home through a grass field. He was in a coma and had difficulty breathing when he was taken to hospital. He died shortly afterwards, and tests indicated that the victim died from a five-pace viper (*Deinagkistrodon acutus*) bite.[14] The name "five-pace" is a local one stemming from the belief that a victim of this viper's bite will only be able to take that many steps before dying.

Snake bite by suicide in real life has also been reported. A 50-year-old man was taken to hospital in India in 2016 after attempting suicide using a Monacled cobra (*Naja kaouthia*). He was in hospital for 5 days, during which he was treated with 20 vials (each around 10 mL) of antivenom.[74] In a completed suicide in the USA in 2019, an 18-year-old male was found in cardiac arrest in an unlocked vehicle in the carpark of a DIY store. This case must have been quite alarming for the police, as closed plastic containers containing various venomous creatures (a hognose viper, a bullfrog, and multiple tarantulas) were found in the car's trunk. The offending snake (also a Monacled cobra) had escaped, but its venom was detected in the man's post-mortem samples, with the highest concentration just under the skin near the puncture marks.[18] The concentration of venom in his serum was 0.158 mg $L^{-1}$, which given that for most snake venoms 0.01 mg $L^{-1}$ is enough to cause symptoms, likely explains the death.[17]

A double homicide in India in 2012 involved the death of an elderly couple (78 and 84 years) from a snake bite. Their son had hired a contract killer over a property dispute and a snake charmer who kidnapped them and persuaded a common cobra (Colubridae) to bite them. Although the killers tried to make it look like an accident, all three men were prosecuted.[75] In an elaborate homicide case in Egypt, a man trained as a snake charmer bought an Egyptian Cobra and forced it to bite his 4-, 6- and 9-year-old female children. All were found dead the next morning with multiple puncture marks and signs of asphyxia and their father was prosecuted.[76]

Despite all the dangers associated with snake venom, it has become a substance of recreational use in some countries. For instance, in India in 2021, a 19-year-old male spent a week in hospital after being deliberately bitten by an unknown snake on the tongue in an attempt to get high. Reportedly, a low dose of venom can bring on a sense of wellbeing, giddiness, drowsiness and euphoria. The mechanism for these effects is not yet understood, and is made more difficult to study by the fact that snake venom abusers have often also used other substances such as cannabis.[77]

For those not keen on being bitten, extracted snake venom (usually illegally trafficked[78]) can be injected with a needle instead, ostensibly for pain relief and sexual desire. The venom can also be dried and made into pills and powders, then added to drinks.[79] The pain relief that some users feel has made buying snake venom a cheaper and longer lasting alternative to opioids. In India in 2018, a 33-year-old male presented to an addiction centre having started using snake bites (possibly from a cobra) regularly instead of opium.[80] Somewhat confusingly, the term 'cobra poisoning' actually refers to drinking a mix of tobacco, solvents and rodenticides, which caused two prisoners in Lithuania to be hospitalised in 2020.[81]

### 9.7   CATCHING THE CHEMICAL CULPRIT

Because venom is a complex biological mixture, most of the techniques we have met so far in this book, can't be used by toxicologists to detect venom in body fluids. An exception is **immunoassay** (see Chapter 6),[13,18,82] but this is only a presumptive technique and can produce false-positives, so cannot give definitive proof for a criminal case. In an emergency hospital setting, it is much more useful because it's so quick and can help medical staff make a treatment decision. However, if the wound has been washed or cleaned before the bite victim arrives at the hospital the test may not be accurate.[15] A new lateral flow test (similar to those used to test for COVID-19) has been developed to distinguish between cobra and viper bites in Taiwan, helping doctors to choose the correct antivenom.[17]

If the identity of the offending snake is unknown, biological methods like DNA profiling of any venom in the blood can give the exact species, but this is time consuming and does not give an indication of venom dose. Another technique is histology – the microscopic examination of changes to cells – which would often be part of a post-mortem examination.[83] Some chemical methods can be used to detect specific molecules or markers in venom.[11] These molecules are typically peptides (strings of linked amino acids), which can be analysed by **liquid chromatography** with **time-of-fight mass spectrometry (TOF-MS)**.[13,78] This works in a similar way to **LC-MS**, which we met in Chapter 2, but instead of looking at the characteristic patterns or ratios the ions make, we fire them down a tube and time how long it takes each one to reach the end. The larger ions are slower than the smaller ones, and each ion has a specific time and mass that we can compare to a library. In Canada in 2014, a two-year-old girl was found deceased with light bruising and a number of small abrasions, after visiting a family friend who kept snakes. SBE was not immediately suspected, but when a post-mortem blood sample was tested using **TOF-MS** and the results compared against a library, there was a match to rattlesnake (*Crotalus*) venom.[8]

## 9.8   CASE CLOSED

Shakespeare bases Cleopatra's death on historical sources, such as the one by Plutarch who lived in the 1st Century CE. Given that Cleopatra died in 30 BCE,[84] he was not born until 75 years after her death,[24] so his records could hardly be called "contemporaneous", and were probably hearsay. The importance of only relying on notes you took at the time of the analysis, and not reporting things you have been told as 'fact' is drummed into every forensic science student.[85]

The story of Cleopatra's dying from the bite of an Egyptian cobra is poetic and intriguing, given these snakes were a symbol of power and divinity – they were thought to deify their victims, and were associated with the Egyptian goddess of healing and magic, Isis.[24,65] But from the angle of a forensic scientist there are many issues with it. First, Shakespeare tells us there were at least two snakes at the scene (which both scarper from an apparently sealed room).[59] Given that Egyptian cobras can be 2 m in length and weigh 9 kg, it seems unlikely that a single convincing 'fig' basket could hold them.[6] Cleopatra's description of the experience of being bitten as "As sweet as balm, as soft as air, as gentle" also does not really fit with the intense pain we know SBE can cause. Then there is the speed of her death, which happens over about 70 words between the first bite and death (about a minute depending on how ostentatious the actress is delivering Cleopatra's final lines).[59] As we have seen, most deaths from SBE by an Egyptian cobra take at least an hour.[24]

Although not the domain of a forensic scientist, there are also some problems with the eyewitness accounts in the play. When

Dolabella (one of Octavius Caesar's soldiers) enters the scene, he asks a guard what has happened. The first guard tries to pin the death on the man with the basket of figs, even though a few lines earlier, we know the guard saw Charmian (one of Cleopatra's ladies-in-waiting) get bitten[59]

| | |
|---|---|
| 1st GUARD: | Caesar hath sent— |
| CHARMIAN: | Too slow a messenger |
| | [Applies an asp] |
| | O, come apace, dispatch, I partly feel thee. |

Act V, scene II
*The Tragedy of Antony & Cleopatra* by William Shakespeare,[40] 1606–7

Because of some of these inconsistencies, rival theories have sprung up about Cleopatra's death. One is that she deliberately drank a mixture of hemlock (Chapter 3), wolfsbane (Chapter 5) and opium (Chapter 2) instead.[86] This idea stems from the notion that the Queen was an experimenter with poisons (on other people, not herself[59]) such as strychnine (Chapter 8), belladonna (Chapter 4) and henbane (Chapter 3).[5,87] Another theory combines the two, suggesting Cleopatra took a high dose of opium as a sedative before being bitten by the cobra,[84] which might explain her dreamy experience of SBE. There are others who believe her death was actually murder by the ruthless Octavius Caesar to remove her political influence from Egypt. In this scenario, she was killed another way (possibly also with poison), then fake fang marks were created on her body (with a brooch pin for example[88]), and the evidence removed from the scene. Octavius then spun the 'romantic' story of the cobra suicide to the people of Egypt.[89]

Whichever method was used to kill Cleopatra, the image of her calm and serene beauty even during what should have been an agonising death, lives on in popular culture.

## REFERENCES

1. W. Shakespeare, *Antony & Cleopatra*, Cambridge University Press, Cambridge, 1950.
2. J. M. Diniz, A. Oliveira Carvalho, R. M. S. Meirelles and H. C. Castro, *Creat. Educ.*, 2021, **12**, 1160–1168.

3. D. A. Warrell, *Saudi Med. J.*, 1997, **18**, 447–452.

4. T. Hargreaves, *Poisons and Poisonings: Death by Stealth*, RSC Publishing, Cambridge, 2017.

5. F. Inkwright, *Botanical Curses and Poisons*, Liminal 11, London, 2021.

6. K. Harkup, *Death by Shakespeare: Snakebites, Stabbings and Broken Hearts*, Bloomsbury, London, 2020.

7. R. M. Renneboog, in *Principles of Chemistry*, ed. D. R. Franceschetti, Salem Press, Ipswich, MA, 2016, ch. 123, pp. 383–386.

8. Q. W. T. Chan, J. Rogalski, K.-M. Moon and L. J. Foster, *Forensic Sci. Int.*, 2021, **323**, 110820.

9. H. Fry and A. Rutherford, in *The Venomous Vendetta*, BBC, London, 2021.

10. *Clarke's Analysis of Drugs and Poisons*, ed. A. C. Moffatt, D. Osselton and B. Widdop, Pharmaceutical Press, London, 4th edn, 2011.

11. *Snake Venoms and Envenomation: Modern Trends and Future Prospects*, ed. Y. Utkin and A. V. Krivoshein, Nova Publishers, New York, 2016.

12. *Venoms: Sources, Toxicity and Therapeutic Uses*, ed. J. Gjersoe and S. Hundstad, Nova Science Publishers, New York, 2010.

13. A. Feola, G. L. Marella, A. Carfora, B. Della Pietra, P. Zangani and C. P. Campobasso, *Toxins*, 2020, **12**, 699.

14. F. Huang, S. Zhao, F. Tong, Y. Liang, J. M. Le Grange, W. Kuang and Y. Zhou, *J. Forensic Sci.*, 2021, **66**, 786–792.

15. Naja haje, http://www.toxinology.com/fusebox.cfm?fuseaction=main.snakes.display&id=SN0181, accessed April 2022.

16. J. L. Tingle and T. Slimani, *J. North Afr. Stud.*, 2017, **22**, 560–577.

17. J.-H. Lin, W.-C. Sung, J.-W. Liao and D.-Z. Hung, *Toxins*, 2020, **12**, 572.

18. D. Paniagua, K. Crowns, M. Montonera, A. Wertheimer, A. Alagón and L. Boyer, *Forensic Toxicol.*, 2020, **38**, 523–528.

19. F. J. Joubert and N. Taljaard, *Eur. J. Biochem.*, 1978, **90**, 359–367.

20. J. Keogh, *Pharmacology*, McGraw-Hill, New York, 2010.

21. M. A. Dkhil, S. Al-Quraishy, H. F. Abdel Razik, A. M. Aref, M. S. Othman and A. E. Abdel Moneim, *Pak. J. Zool.*, 2014, **46**, 1719–1730.

22. A. Abdel-Aal and A. A. Abdel-Saset, *Sci. J. King Faisal Univ.*, 2010, **11**, 167–182.
23. T. Stone and G. Darlington, *Pills, Potions, Poisons: How Drugs Work*, Oxford University Press, Oxford, 2000.
24. R. Girling, in *The Sunday Times*, London, 2004.
25. E. D. Brodie III, *Curr. Biol.*, 2018, **28**, R1221–R1242.
26. S. Al-Quraishy, M. A. Dkhil and A. E. Abdel Moneim, *J. Venomous Anim. Toxins Incl. Trop. Dis.*, 2014, **20**, 42.
27. T. R. Rahmy, *J. Venomous Anim. Toxins*, 2001, 7, 85–112.
28. F. A. Adamude, E. J. Dingwoke, M. S. Abubakar, S. Ibrahim, G. Mohamed, A. Klein and A. B. Sallau, *Toxicon*, 2021, **197**, 24–32.
29. S. Vasudev, V. S. More, K. S. Ananthraju and S. S. More, *J. Ayurveda Integr. Med.*, 2021, **12**, 458–464.
30. D. A. Warrell, H. J. Barnes and M. F. Piburn, *Trans. R. Soc. Trop. Med. Hyg.*, 1976, **70**, 78–79.
31. I. Bolon, A. M. Durso, S. Botero Mesa, N. Ray, G. Alcoba, F. Chappuis and R. Ruiz de Castañeda, *PLoS One*, 2020, **15**, e0229989.
32. Y. K. Gupta and S. S. Peshin, *Toxicol. Int.*, 2012, **19**, 89–99.
33. M. Chamberlin, *Old Wives' Tales: The History of Remedies, Charms and Spells*, The History Press, Cheltenham, 2020.
34. E. Schioldann, M. A. Mahmood, M. M. Kyaw, D. Halliday, K. T. Thwin, N. N. Chit, R. Cumming, D. Bacon, S. Alfred, J. White, D. Warrell and C. A. Peh, *PLoS Neglected Trop. Dis.*, 2018, **12**, e0006299.
35. S. P. Bush, *Ann. Emerg. Med.*, 2004, **43**, 187–188.
36. A. Dey and J. N. De, *Afr. J. Tradit., Complementary Altern. Med.*, 2011, **9**, 153–174.
37. Y. Padma, N. Sarojinidevi, K. Venkata Ratnam, G. Tirupati Reddy and R. R. Venkata Raju, *J. Med. Plants Stud.*, 2016, **4**, 52–56.
38. M. Brown, *Death in the Garden: Poisonous Plants & Their Use Throughout History*, Pen & Sword Books Ltd, Barnsley, 2018.
39. M. Ovid, *Metamorphoses*, Penguin, London, 2004.
40. W. Shakespeare, *Macbeth*, Cambridge University Press, Cambridge, 1960.
41. C. J. Reading, *J. Accid. Emerg. Med.*, 1996, **13**, 346–351.
42. T. Lamb, D. Stewart, D. A. Warrell, D. G. Lalloo, P. Jagpal, D. Jones, R. Thanacoody, L. A. Gray and M. Eddleston, *Clin. Toxicol.*, 2021, **59**, 992–1001.

43. R. Kipling, *The Jungle Book*, Harper Collins, London, 2010.
44. J. K. Rowling, *Harry Potter and the Philosopher's Stone*, Bloomsbury, London, 1997.
45. J. K. Rowling, *Harry Potter and the Chamber of Secrets*, Bloomsbury, London, 1998.
46. J. K. Rowling, *Harry Potter and the Goblet of Fire*, Bloomsbury, London, 2000.
47. J. K. Rowling, *Harry Potter and the Order of the Phoenix*, Bloomsbury, London, 2003.
48. J. K. Rowling, *Harry Potter and the Deathly Hallows*, Bloomsbury, London, 2007.
49. British Library, *Harry Potter: A History of Magic*, Bloomsbury, London, 2017.
50. R. Highfield, *The Science of Harry Potter: How Magic Really Works*, Headline Book Publishing, London, 2002.
51. A. Chevallier, *Encyclopedia of Medicinal Plants*, DK Publishing, St Leonards, 2001.
52. Medical Economics Company, *PDR for Herbal Medicines*, Medical Economics Company, Montvale, NJ, 1998.
53. D. Frohne and H. J. Pfänder, *Poisonous Plants: A Handbook for Doctors, Pharmacists, Toxicologists, Biologists and Veterinarians*, Manson Publishing, London, 2nd edn, 2004.
54. S. T. Dietz, *The Complete Language of Flowers: A Definitive and Illustrated History*, Wellfleet Press, New York, 2020.
55. M. B. Pucca, C. Knudsen, I. S. Oliveira, C. Rimbault, F. A. Cerni, F. H. Wen, J. Sachett, M. A. Sartim, A. H. Laustsen and W. M. Monteiro, *Toxins*, 2020, **12**, 668.
56. J. J. Russell, A. Schoenbrunner and J. E. Janis, *Plast. Reconstr. Surg.*, 2021, **9**, e3506.
57. L. Snicket, *A Serious of Unfortunate Events Book #2: The Reptile Room*, Harper Collins, New York, 1999.
58. D. R. Dalessandro, J. Heffernan and S. Guiterrez, *Snakes on a Plane Screenplay*, New Line Cinema, Burbank, CA, 2006.
59. A. M. Kinghorn, *English Stud.*, 1994, **75**, 104–109.
60. S. M. Kane, *Appalachian J.*, 1974, **1**, 255–262.
61. *The New Oxford Annotated Bible (New Revised Standard Version)*, ed. M. D. Coogan, M. Z. Brettler, C. A. Newsom and P. Perkins, Oxford University Press, Oxford, 2006.
62. J. Clabeaux, *Cathol. Biblical Q.*, 2005, **67**, 604–610.
63. S. Mifsud, in *Times Malta*, Allied Newspapers Ltd, Birkirkara, 2014.

64. S. C. Greene, J. Folt, K. Wyatt and N. P. Brandehoff, *Am. J. Emerg. Med.*, 2021, **45**, 309–316.

65. D. Morgan, *Snakes in Myth, Magic and History*, Praeger, Westport, CT, 2008.

66. S. Senthilkumaran, N. Karthikeyan, N. N. Jena and P. Thirumalaikolundusubramanian, *Med. Sci. Law*, 2017, **57**, 158.

67. A. Lafnoune, S.-Y. Lee, J.-Y. Heo, I. Gourja, B. Darkaoui, Z. Abdelkafi-Koubaa, F. Chgoury, K. Daoudi, S. Chakir, R. Cadi, K. Mounaji, N. Srairi-Abid, N. Marrakchi, D. Shum, H.-R. Seo and N. Oukkache, *Toxins*, 2021, **13**, 402.

68. M. Thompson, in *Pharmaceutical Journal*, Royal Pharmaceutical Society, London, 2009.

69. S. Mohamed, A. El Amir, L. S. E. L. din Shaker, A. Elfeky and W. S. Nabil, *Egypt. J. Chem.*, 2022, **65**, 739–750.

70. A. Dematei, J. B. Nunes, D. C. Moreira, J. A. Jesus, M. D. Laurenti, A. C. A. Mengarda, M. S. Vieira, C. P. do Amaral, M. M. Domingues, J. de Moraes, L. F. D. Passero, G. Brand, L. J. Bessa, R. Wimmer, S. A. S. Kuckelhaus, A. M. Tomás, N. C. Santos, A. Plácido, P. Eaton and J. R. S. A. Leite, *J. Nat. Prod.*, 2021, **84**, 1787–1798.

71. P. Lazarovici, C. Marcinkiewicz and P. I. Lelkes, *Toxins*, 2019, **11**, 303.

72. Y. Guo, Y. Wang, X. Zhao, X. Li, Q. Wang, W. Zhong, K. Mequanint, R. Zhan, M. Xing and G. Luo, *Sci. Adv.*, 2021, 7, eabf9635.

73. N. Ibrahim and N. Farid, *J. Appl. Sci. Res.*, 2009, **5**, 1223–1229.

74. S. Mallik, S. R. Singh, M. K. Mohanty and N. Padhy, *Med. Sci. Law*, 2016, **56**, 264–266.

75. V. N. Ambade, J. L. Borkar and S. K. Meshram, *Med. Sci. Law*, 2011, **52**, 40–43.

76. M. G. Paulis and A. L. Faheem, *J. Forensic Sci.*, 2016, **61**, 559–561.

77. D. Ram, *Indian J. Physiol. Pharmacol.*, 2022, **65**, 242–244.

78. K. L. Verma, A. P. Singh and S. Sinha, *Forensic Toxicol.*, 2018, **36**, 537–539.

79. S. Das, P. Barnwal, T. Maiti, A. Ramasamy, S. Mondal and D. Babu, *Subst. Use Misuse*, 2017, **52**, 1104–1109.

80. A. Mehra, D. Basu and S. Grover, *Indian J. Psychol. Med.*, 2018, **40**, 269–271.

81. M. Perminas, J. Surkus, R. Leksiene and A. Stankuviene, *40th International Congress of the European Association of Poisons Centres and Clinical Toxicologists (EAPCCT)*, Tallinn, Estonia, 2020, p. 527.

82. G. Brunda, R. B. Sashidhar and R. K. Sarin, *Toxicon*, 2006, **48**, 183–194.

83. *Forensic Pathology*, ed. M. M. Houck, Academic Press, Amsterdam, 2017.

84. A. M. Rosso, *Toxicol. Rep.*, 2021, **8**, 676–695.

85. A. R. W. Jackson and J. Jackson, *Forensic Science*, 4th edn, Pearson, Harlow, 2018.

86. Poison, not snake, killed Cleopatra, scholar says, https://web.archive.org/web/20120912193217/http://articles.cnn.com/2010-06-30/world/cleopatra.suicide_1_cleopatra-snake-cobra?_s=PM:WORLD, accessed March 2022.

87. Anonymous, *The Poison Garden*, The Alnwick Garden, Alnwick, 2005.

88. B. Hubbard, *Poison: The History of Potions, Powders and Murderous Practitioners*, Welbeck, London, 2020.

89. G. Tsoucalas and M. Sgantzos, in *Toxicology in Antiquity*, ed. P. Wexler, Academic Press, New York, 2nd edn, 2019, ch. 4, pp. 83–92.

# CHAPTER 10

# Nature's Toxic Gift

If you see a term that's **bold** it's defined in the Glossary. Only the first time that the word appears in the chapter will it be indicated in this way.

---

**Case History: Crime of Protection**

Mrs Watkins went into the pantry and took out a liqueur-glass. She poured a little sloe gin into it, then she put down the bottle and left the pantry. She went into the children's dark-room—they were allowed that for their photography.

She still had the glass in her hand. There was a bottle on the highest shelf. She took it down and measured it carefully with her eye. The children's manual of photography and the medical dictionary in Henry's dressing-room had been a great help to her.

She poured out into the deep red of the sloe gin some of the contents of the bottle; it looked very white and harmless and hardly smelt at all.

*The Liqueur Glass* by Phyllis Bottome,[1] 1915

---

Poisonous Tales: A Forensic Examination of Poisons in Fiction
By Hilary Hamnett
© Hilary Hamnett 2023
Published by the Royal Society of Chemistry, www.rsc.org

## 10.1   THE INVESTIGATION

Mrs Watkins has been planning this murder for some time. Fed up with her controlling and abusive husband, Henry Watkins, she plots his demise week after week during the sermons in church. Although he is wealthy, her motive is not greed, but a desire to set her two adult children, Hetty and Paul, free of his tyranny. She cleverly waits until the servants are out of the way, then uses some sloe gin that was sent to Henry by his brother, and which he has not yet tasted, to deliver the deadly dose

Then she went back to her husband. "Here it is, Henry," she said. "What a slow woman you are!" he grumbled. "Still I must say you have a steady hand."
She held the full glass towards him and watched him drink it in a gulp. "It tastes damned odd," said Henry thoughtfully. "I don't think I shall take any more of it."
Mrs Watkins did not answer; she took up the liqueur-glass and went back into the pantry. She took out another glass, filled it with sloe gin, drank it, and put it on the pantry table. The first glass she slipped up her long sleeve and went out into the garden.

*The Liqueur Glass* by Phyllis Bottome,[1] 1915

We see few further details of Henry's death, but the dialogue suggests he dies within minutes. His death was followed by an inquest, which meant everyone in the house was questioned and the liqueur glass examined. Of course, nothing was found, as Mrs Watkins had planted a fake in the pantry and disposed of the real glass (we later find out she smashed it in the garden). The verdict is "death through misadventure" with a doctor declaring poison was the cause. The theory put forward is that Henry took it deliberately, but a verdict of suicide in the 1900s would have needed evidence beyond reasonable doubt. This was partly due to the stigma and shame associated with suicide, and the fact that it was a crime in England until the Suicide Act of 1961, but mainly due to the refusal of the church to allow those dying as a result of suicide to have a Christian burial.

So which poison, found in a photography dark-room, could have brought about such a rapid death? Many of the chemicals used to develop black & white prints during the time the story was written (1915) were poisonous, but each had a different way of acting on the body.

The chemistry behind pre-digital photography was based around silver. A light-sensitive film contained crystals of silver halide (AgBr). When it was exposed to light, it broke down into components, one of which was silver metal. So, the first (or latent) image was created from a tiny amount of silver on the film. In the "development" stage another chemical was added to encourage the silver metal crystals to grow, until the whole image appeared, made of silver (although this would still look black to the human eye). Once enough silver had grown over the film, it was placed in a "stop bath" containing a weak acid such as vinegar. Now to get rid of the unused silver halide, so the film is no longer light-sensitive, a "fixer" was used to dissolve it away. In the final stage, various different "toners" were added to convert the silver metal into a more stable silver salt (to stop it from tarnishing). Depending on the effect that was desired, this could produce pictures that were tinted with colour, such as sepia (brown).

Amongst these various chemicals, some were poisonous because they were corrosive (they burned the skin), such as phenol. But Henry merely comments on the taste of the sloe gin, and doesn't show any symptoms of burning. Others, like hydrochloric acid, smell very pungent, and Mrs Watkins barely notices a smell as she handles the poison. Then there are various metal salts, of barium, lead and mercury. These are indeed poisonous, but their effects, as we will see in Chapter 11, are usually vomiting and stomach pains, rather than sudden death.[2] Other photography chemicals are violently explosive, such as chromium trioxide, which would have been something of a giveaway if added to a drink. **Oxalic acid**, found in rhubarb (*Rheum rhabarbarum*) leaves was used in photography as a toner, and is poisonous, but tends to cause vomiting, convulsions and slow kidney failure,[3,4] and a large dose is needed to cause a quick death.[5,6] Something that was used to clean glass bottles of photography chemicals was potassium bichromate, but this is not soluble in alcohol, so would have been obvious if floating around in the sloe gin.[7]

That leaves us with cyanides, a general name given to molecules that contain the $-CN$ or $-C\equiv N$ unit.[8] Despite being well-known as deadly poisons, several different cyanide compounds were used in black & white photography, until they were finally considered too much of a hazard in the 1980s. Potassium ferricyanide [$K_3Fe(CN)_6$] was used to alter the tones of prints to produce "cyanotypes" or blue prints, with the chemical being sold as "Farmer's solution" or "Farmer's reducer". Unfortunately, it is bright red, and Mrs Watkins describes the poison as white. That leaves the white potassium cyanide (KCN), which smells slightly of bitter almonds and is used as a fixer.

In this case a doctor states at the inquest that the cause of death was poison, but we do not know why he thought that. At the time the story was written there were several tests known for cyanide, and it is possible the author knew of these. One test, which was published in the 1800s describes adding silver nitrate to cyanide, to give a white precipitate.[9] Like many other methods for detecting poisons developed in the 19th Century, it was a simple positive or negative test,[10] and was most likely to have been carried out on the liqueur glass rather than on samples taken from the body (as these make it hard to see the colour change). Toxicology testing may not have been necessary however, as the smell of bitter almonds on a body and having pink skin are signs of cyanide poisoning, which may have convinced the doctor of the cause of death.[11,12] The fact that Mrs Watkins appears not to notice the smell of bitter almonds on the poison may be because it's a genetic skill you have to inherit, which means up to half of us can't smell cyanide.[13]

## 10.2 THE POISON BEHIND THE STORY

### 10.2.1 Cyanide

Cyanides are poisonous because they contain a –CN bond. How toxic they are depends on how strong the bond is, with hydrogen cyanide (HCN, also known as "Prussic acid" or "hydrocyanic acid") having the weakest.[8,13] When the bond breaks, cyanide ions ($CN^-$) are released, which, as we will see, wreak havoc in the body. Salts such as KCN are particularly dangerous as they react with water (such as the water in sloe gin) to make HCN. If the water is then drunk, it releases cyanide ions into the bloodstream.

Despite the demise of black & white photography, cyanides are still used today in industry as insecticides, for hardening of steel, electroplating of metals and mining gold and silver.[12,13] Many plants of the *Prunus* species also naturally harbour the cyanide-containing **amygdalin**, a **cyanogenic glycoside**, particularly in their seeds, pips or stones (*e.g.*, peaches, elderberries, cherries, apples and bitter almonds – not the usual sweet kind), which can be broken down by **enzymes** called "glucosidases" into HCN.[6,14] You may have been told as a child not to swallow an apple pip, with your parents inventing a story about why not, but these folk tales have probably evolved from a poisonous encounter in the past. In reality, whole pips and stones are quite sturdy and usually pass through the digestive system without releasing cyanide. And if a small amount of cyanide does escape, we can naturally metabolise it using the enzyme rhodanase (enzyme names end with "–ase") into a less toxic metabolite called thiocyanate (–SCN).[8]

Another way to make HCN is by burning nitrogen-containing materials, so it's found in vehicle exhausts, cigarette smoke, and the fumes given off during house fires.[15] This is particularly true for fires involving soft furnishings, such as mattresses and sofas, stuffed with **polyurethane** foam, which releases HCN when it burns.[16] After a fatal fire in a department store in the UK in 1979, The Furniture and Furnishings (Fire) (Safety) Regulations 1988 came in, forcing manufacturers to make their polyurethane furnishings flame retardant.

But even if you don't smoke, eat any fruit, work with cyanide or develop photographs as a hobby, cyanide is made by the body as a result of normal metabolism.[13] So, we all have a small amount of cyanide in our blood (around 0.004 mg $L^{-1}$) without suffering any ill-effects.[13] Other poisons like carbon monoxide (Chapter 12) are also present in non-toxic amounts in the body, and this must be taken into consideration when interpreting what toxicology results mean in a case.

Larger doses of cyanide are a different matter and can cause a very rapid death indeed (within 15 to 45 min[17]) because cyanide is a systemic poison that interferes with our ability to use the oxygen we inhale.[18] It does this by hijacking an oxygen-binding enzyme called "cytochrome oxidase", which is a key player in the processing of oxygen in our cells. The iron in this enzyme usually likes to bind to an oxygen ($O_2$) molecule, but when $CN^-$ is loose in the bloodstream it binds to that instead. Within minutes of eating or drinking cyanide, the cells of the body are starved of oxygen, even if the person is still breathing.[19] The body is its own worst enemy in cyanide poisonings as it pumps it around in the blood delivering it to all the cells and organs.[17] This lack of oxygen brings on dizziness and weakness followed by unconsciousness, coma and respiratory failure.[20] It's a kind of chemical strangulation or suffocation.[17,21] Slightly confusingly "cyanosis" means turning blue, but the skin of someone who has been poisoned with cyanide is actually pink, and their blood samples are bright cherry red.[22] If the skin or lips of a cyanide victim has turned blue, it's because of the eventual cardiac collapse and loss of circulation, something we see in other poisonings as well.[23] How much cyanide is fatal depends on how it is taken (with inhalation being the most dangerous) and whether or not the victim seeks help.

There *are* treatments for cyanide poisoning, but as we have seen, large doses act so quickly there is usually little time for antidotes. Cyanide antidotes work either by binding to the $CN^-$ ions themselves to prevent further damage, or by speeding up the body's natural detoxification processes.[6] Drugs such as nitrites work by luring the cyanide ions away from the cytochromes and onto methaemoglobin instead. There is not enough methaemoglobin in the blood normally, so nitrite creates some more by oxidising haemoglobin.[19] Care needs to be taken with the dose, as this itself can be dangerous, and consuming large amounts of sodium nitrite ($NaNO_2$) has become a suicide method that forensic toxicologists are seeing more and more often.[24,25] This is probably because nitrite can be freely purchased on the internet, and this method has gained attention following some celebrity suicides. Toxicologists can detect methaemoglobin (and nitrite poisonings) using the same technique we use to detect carbon monoxide (Chapter 12). Another cure is to use hydroxocobalamin (vitamin B12$_a$), which reacts with cyanide to make cyanocobalamin, which is removed in the urine.[13,19,26] Another treatment is to give the person sodium thiosulphate ($NaS_2O_3$), which converts cyanide into the less toxic metabolite thiocyanate.[27]

Plants containing cyanogenic glycosides have long been used in Ayurvedic medicine, for diseases including parkinsonism. For example, the velvet bean (*Mucuna pruriens*) is thought to be active against snake venom (see Chapter 9) and if applied as a paste on scorpion stings, is thought to absorb the poison.[28] In folk medicine, the milk of peach (*Prunus persica*) kernels was applied to the forehead to promote sleep, or boiled in vinegar and applied to the head to cure baldness.[29] The dried leaves of the Cherry laurel (*Prunus laurocerasus*) have been used as a tonic for the stomach, and as an anti-irritant, and in homeopathy for coughs.[30] Cherry laurel water was also used as a mild narcotic known as *Aqua Laurocerasi*.[31] The Cherry laurel looks very similar to its edible cousin the Bay tree (*Laurus nobilis*) and its leaves can be mistaken for Bay leaves in cooking (see Figure 7.4, left).[29]

Bitter almond (*Prunus dulcis* var. *amara*) is no longer used in Western medicine, but in the past was used as almond water to treat vomiting and nausea.[32] It is still a popular remedy in Traditional Chinese Medicine (TCM) called *Xing Ren* for coughs and asthma.[33] However, care must be taken with the dose as swallowing 6–10 almonds may cause serious poisoning while 50–60 may cause death.[34] The bitter taste puts most people off accidentally eating them, but as we saw earlier with the smell of cyanide, not everyone notices the taste.[29]

### 10.2.2 Sloe

Sloe was only the vehicle for the poison in our case history – but has been known to cause infection when gardeners scratch themselves on its spines.[35] Sloe is a common name for blackthorn (*Prunus spinosa*), also known as "Mother of the wood", "Hedge Picks" and "Wishing thorn".[36] Blackthorn is a familiar sight in winter in the hedges of the UK as a short (~4 m high) shrub. It has white flowers and spherical very sour blue fruit called "sloes" (Figure 10.1), which are used to make sloe gin.[29] Making sloe gin doesn't involving any fermenting or distillation – the recipe starts with ready-made gin, which is seeped with the berries of the blackthorn and sugar for several months to add colour and flavour.[37] Originally, sloes were added to mask the unpleasant flavour of cheap gins, but sloe gins are now enjoying something of a hipster revival.

The blackthorn plant also has a history in folk medicine; sloe juice was used as gargle preparation for a sore throat, the flowers (Figure 10.1, left) treated common colds, and in homeopathy blackthorn was used for nervous headaches.[38]

**Figure 10.1**    (Left) The flowers and (right) fruit of the blackthorn (*Prunus spinosa*) or sloe plant, taken in Lincoln.

## 10.3   POISONOUS PLOTS

Perhaps the most familiar use of cyanide in fiction is in spy plots where capsules of cyanide salts are used by secret agents as a suicide method when they are captured. This may be more fiction than fact, as when the body of an SS soldier who died in 1944 was unearthed with a dozen vials of 'poison', none of them actually contained cyanide.[39] Still, cyanide makes many cameos in the *James Bond* films and books including the first one, *Dr No.* (1962) where it was hidden in a cigarette (a tactic from real-life crime too[40]), right up to the more recent *Skyfall* (2012).[41] Foaming at the mouth after ingesting cyanide is often depicted in films, including in *Captain America: The First Avenger* (2011) but this is more for dramatic effect, and is not a common symptom of eating cyanide salts.

The poisoning of Hamlet Snr, which we explored in Chapter 3, is only the first of five deaths in *Hamlet* (1600) involving poison.[42] Queen Gertrude (Hamlet Jnr's mother) dies after drinking from a poisoned cup, and cyanide is a likely candidate for that one. It is also another possibility for the fast-acting poison that Romeo drinks in *Romeo & Juliet* (1594).[43,44] At the end of the play, Juliet kills herself by kissing Romeo in the hope that enough poison remains on his lips to take her life as well. It's possible that cyanide could be passed between people *via* the lips, by the second

**Figure 10.2**    Dissolution of Partnership, by Hablot Knight Browne.

person breathing in HCN gas, and in fact extreme care must be taken by anyone performing mouth-to-mouth resuscitation on a cyanide poisoning victim.[45] Cyanide salts were not extracted from almonds until around 1800,[46] long after Shakespeare's day, but as we have seen, there were plenty of plant sources of cyanide around.[47]

There are also multiple murders in Charles Dickens' novel *Martin Chuzzlewit* (1843) (Figure 10.2), the first of which was the poisoning of Anthony Chuzzlewit by his greedy son and business partner, Jonas, most likely using **strychnine** (Chapter 8).

A second (non-poisoning) murder follows as Jonas tries to cover his tracks, but he is eventually arrested. In the coach on the way to prison he takes his own life with a different poison

Happening to pass a fruiterer's on their way; the door of which was open, though the shop was by this time shut; one of them remarked how faint the peaches smelled. The other assented at the moment, but presently stooped down in quick alarm, and looked at the prisoner.

'Stop the coach! He has poisoned himself! The smell comes from this bottle in his hand!' The hand had shut upon it tight. With that rigidity of grasp with which no living man, in the full strength and energy of life, can clutch a prize he has won. They dragged him out into the dark street; but jury, judge, and hangman, could have done no more, and could do nothing now. Dead, dead, dead.

Chapter 51
*The Life and Adventures of Martin Chuzzlewit* by Charles Dickens,[12] 1843

We know that the stones of peaches contain amygdalin, so it's possible the poison was made from crushed peach kernels similar to the one the Ancient Egyptians made their prisoners drink (a distillate of crushed peach kernels in water, called the "penalty of the peach").[48] Cyanide is also the most likely candidate from the speed of Jonas' death; again the narrative suggests he is dead within minutes. If you were to try to eat raw peach kernels to get the same effect, you would need >20 and possibly as many as 150 depending on the fruit (the amount of cyanide in each kernel varies widely).[49]

The smell of peaches and almonds also feature in the novel *The Rubicon* (1894) by Edward Benson. The heroine, Eva, takes her own life with some Prussic acid she finds in a laboratory. We don't see the death itself but Eva describes how the "harmless-looking liquid" reminded her of the almond icing on top of wedding cakes.[50]

In the graphic novel *V for Vendetta* (1989), a modern take on George Orwell's *1984*, about an authoritarian regime in England in 1997, the hero insurgent who is known only by the alias "V" sets about trying to kill as many of those in power as he can. His tactics involve blowing up famous London landmarks and targeting individuals, such as the Catholic bishop Anthony Lilliman. Lilliman was the chaplain at the prison camp where V was held, and the chaplain knew of the experiments conducted on V and the other prisoners. V lures him into a trap and forces him to eat a communion wafer laced with cyanide.[51] We only see the body of the bishop and not the death itself, but a wafer could be a vehicle for a fatal dose of cyanide.

In the 2008 episode of *Doctor Who*, called the *Unicorn and the Wasp* about the murder mystery author Agatha Christie (herself a frequent user of cyanide in her poisonous plots) there are three deaths. Two are down to stings from a giant alien wasp, one is killed by falling masonry, and then there is the attempted murder of the Doctor with cyanide

*[The butler, Greeves, enters the sitting room with a tray of drinks]*
GREEVES: Your drinks, ladies, Doctor.
DOCTOR: Very good, Greeves.
*[Greeves leaves and the Doctor takes a sip]*
DOCTOR: No. Something's inhibiting my enzymes. Argh! I've been poisoned.
*[The Doctor is nearly doubled up in pain]*
DONNA: What do we do? What do we do?
*[Agatha sniffs his drink]*
AGATHA: Bitter almonds. It's cyanide. Sparkling cyanide.

Season 4, episode 7
*The Unicorn and The Wasp* directed by Graeme Harper, 2008

Yet this is a fantasy plot and the Doctor, himself an alien with a different set of enzymes to humans, escapes death with a combination of ginger beer and anchovies.

Two more recent on-screen poisonings with unnamed substances involved poisoned coffee in Quentin Tarantino's film *Hateful 8* (2015), and the murder of King Joffrey in season 4 of *Game of Thrones* (2014) after drinking poisoned wine. In Koushun Takami's book *Battle Royale* (1999) a group of 42 ninth-grade (14- and 15-year-old) school students is taken to a deserted island where they are forced to kill one another. One of the female students, Yuka Nakagawa, accidentally eats some poison-laced soup. The poison is described as a "half-transparent powder" and is intended for another student

Yuka dropped the dish and the stew splashed against the floor with a crashing sound. Everyone looked over at her. Yuka held onto her throat and coughed out the stew she had just swallowed...Yuka balled up on her side and coughed up blood again. Her tan face turned pale then blue. Red foam spilled out the side of her mouth.

Chapter 62, p. 455
*Battle Royale* by Koushun Takami (trans. Yuji Oniki),[52] 1999

Fans of all three stories believe that cyanide was the culprit, but although some of the symptoms point to cyanide, the coughing up of blood is not a well-known one and is likely to be for dramatic effect.

## 10.4 MODERN MEDICAL USES

Given the frightening toxicity of cyanide compounds, you may be surprised to discover that one of them, an intense blue dye called Prussian blue [$Fe_7(CN)_{18}$], is a useful antidote against other poisons. It can be taken as a pill by those exposed to radioactive caesium and non-radioactive thallium metals. It works by trapping the metals in the digestive system, preventing them from being absorbed.

The cyanide-containing sodium nitroprusside ($Na_2[Fe(CN)_5NO]$), has been used for emergency conditions such as heart failure for decades. The drug itself is not active, but it breaks down in the body to nitric oxide (NO), which can rapidly lower blood pressure.[13] Drugs that are inactive initially but turn into something useful in the body are known as pro-drugs (see Chapter 3). Unfortunately, the breakdown of sodium nitroprusside also releases $CN^-$ ions into the bloodstream, which can lead to toxic side-effects.

Naturally occurring cyanide sources have also been used in modern medicine. The controversial alternative cancer cure **laetrile** is used as a label for several different but related substances: amygdalin, mandelonitrile or D-mandelonitrile-β-glucuronide, all of which can release HCN when ingested. Some of these compounds are extracted from fruit pips and stones, and others are made in the laboratory.[6,53,54] Laetrile is controversial because various studies have shown it to be at best ineffective against cancer, and at worst highly toxic,[55] including to a 4-year-old who swallowed 12 tablets.[13] The toxicity to the toddler was due to cyanide itself, but there have also been reports of biological and other contaminants in laetrile.[56] In another case, a 17-year-old ingested three ampules of laetrile and died in hospital within 24 h.[57]

Mandelonitrile-β-glucuronide is a member of a group of chemicals called "glucuronides". Glucuronides are very familiar to toxicologists, as they are made during the metabolism (breaking down) of all sorts of drugs in the body. Metabolism of drugs by the body usually aims to make them less harmful and more water soluble so they can be passed in the urine. We can think of metabolism as happening in two stages. In the first phase, small changes are made to the drug or poison's chemical structure – a few atoms are added or removed, creating a chemical 'hook'. In drugs such as opiates (*e.g.*, **morphine**) or benzodiazepines (*e.g.*, **diazepam**) these reactions are just preparing the ground for the second phase when larger molecules attach themselves to the drug or poison with the help of **enzymes**. The most common attachment is the sugar glucuronic acid, and when this is added to a drug it becomes a "glucuronide". The best place to look for glucuronides is in urine samples, although they can also be found in blood samples in some cases.

Forensic toxicologists do not always investigate whether glucuronides are present, but they can be useful in situations where there has been a long delay before the samples were taken, for example in a drug-facilitated sexual assault. Even if all the 'parent' drug has gone from the complainant's system, there may still be some glucuronides in their urine that can shed light on what happened. In a **heroin** overdose death for instance, a high amount of parent drug and a small amount of glucuronide in the urine suggests the person died quickly after taking the heroin. If they injected the heroin, this is known as a "hotshot" death.[58]

### 10.5 MODERN TOXICOLOGY CASES

The glamorous spy plots and intriguing murders of fiction are all very well, but forensic toxicologists are actually most likely to encounter cyanide in fire deaths. For instance, in the state of Victoria in Australia between 1992 and 1998 there were 178 fire deaths, of which 86 tested positive for cyanide.[59] In Sweden between 1992 and 2009 there were 2303 fire fatalities, of which one-third had a toxic level of cyanide in the blood.[60] Finally, in Poland between 1995 and 2011, there were 285 fatal fire victims of which 59% were positive for cyanide.[61] However, whether or not cyanide is the sole cause of death in fire cases is complicated by the victims also breathing in carbon monoxide (Chapter 12).[62] Approximately 35% of those presenting to A&E after a fire will have toxic levels of cyanide ($>0.2$ mg L$^{-1}$ [45]) on arrival,[63] but cyanide may not be noticed as contributing to inhalation injuries.[64] This is possibly because carbon monoxide poisoning is much more widely publicised, but also because cyanide poisoning is harder to recognise.[65] There is also no quick-and-easy test for cyanide poisoning, like there is for carbon monoxide,[66] and a cyanide test can take hours or even days.[58]

If the fire victim does not survive, toxicology testing on severely burnt bodies can be difficult, as there may not be any blood left to sample at the autopsy.[67] Spleen, an organ containing a lot of red blood cells, can be a useful specimen in these situations (see Chapter 12).[61] Even if there is a blood sample, cyanide is also not always tested for in toxicology laboratories, particularly if another drug or poison has already been found at a high level.

We saw in Chapter 8 how strychnine poisoning can be misdiagnosed as a tetanus infection, because the symptoms look so similar. The same is true for cyanide and diabetes. Cyanide poisoning can cause harmful substances called "ketones" to build up in the body eventually lowering the pH of the blood to dangerous levels (known as "acidosis"), the same effect as seen in uncontrolled diabetes.[33,68] In a case in Taiwan in 2020, a 37-year-old woman was found unconscious in a park and sent to A&E. Discovering her low blood pH, doctors thought she was diabetic and treated her with insulin. It was only when she woke up 12 h later and admitted eating a bag of bitter almonds in a suicide attempt, that they realised cyanide was the cause of her symptoms.[69] A 30-year-old woman injected herself under the skin with cyanide in Spain in 2005 and also suffered acidosis, but her case was more clear-cut as she was found with the hypodermic syringes containing cyanide dissolved in ethanol.[62]

The sale and possession of cyanide compounds requires a licence in many countries, and, similar to strychnine, it is regulated under the Poisons Act 1972 in the UK. This means that suicides involving cyanide salts are more likely to be seen in people with access to chemicals through their work or place of study. In one case in Brazil in 2018, a 56-year-old man who worked as a goldsmith was found dead at home next to a plastic bottle of what appeared to be a cyanide salt. In a second case, an 18-year-old female student on a chemistry course in Brazil was found unconscious on campus. Her phone contained selfies of her holding a flask of KCN (potassium cyanide). The levels of cyanide in these two cases were 31 and 27 mg $L^{-1}$ in post-mortem blood,[70] which are very high considering fatal concentrations are usually between 1 and 3 mg $L^{-1}$.[71]

In the second Brazilian case, technology was a help to the police in gathering evidence, but it has also given people widespread access to websites that sell poisons[13] and sites that share suicide methods. "Suicide kits" containing KCN or its precursors can be bought on Ebay® for example, and in 2021 in Poland, there were 4 fatal intoxications in 3 months related to internet shopping. In one case, a 44-year-old male chemist was taken to ICU after synthesising KCN from a kit he bought online. He died later, but interestingly no cyanide was detected in any of his post-mortem samples.[72] This is likely due to the antidote he was given in hospital, which binds free cyanide into a large complex and passes it through the urine, making it invisible to forensic toxicologists.[73]

Another cause of fatal poisoning is cassava (*Manihot esculenta*), a root vegetable that contains the cyanogenic glucoside **linamarin**. It can be eaten safely if the cassava flour is processed first either by wetting then drying the paste over a period of hours, or by fermenting cassava root to allow the HCN to evaporate. Skipping this step can have fatal consequences, as was seen in Nigeria in 1992 when three members of the same family died after eating a cassava meal.[74] There is also the problem of the longer term chronic poisoning disease called *konzo*, which causes paralysis of the legs, dizziness, headaches and vomiting.[75]

A rare homicide involving cyanide happened in Iran in 2012 when a 29-year-old man was given a poisoned glass of beer. He immediately started vomiting and became dizzy. He was taken to hospital but died shortly afterwards, and cyanide was found in his post-mortem samples including his stomach contents, suggesting he had drunk the poison.[67] The stomach is a good place to look for cyanide during the autopsy as gastric burns are often seen in these cases.[76] However, taking stomach contents samples is a risky business, as the cyanide salt will react with the stomach acid and release HCN gas when the stomach is opened.[61,77]

## 10.6   CATCHING THE CHEMICAL CULPRIT

**Gas chromatography** (GC) is often used to detect cyanide in biological samples,[78] but as we saw in Chapter 4, drugs and poisons need to be extracted from biological fluids before they can be analysed by sensitive instruments. A commonly used extraction method for cyanide (and also alcohol) is **headspace extraction**.[72,79] This involves placing the blood in a tightly sealed tall glass tube and heating it gently, at about 50 °C. Gradually some of the HCN evaporates from the liquid into the air (or "headspace") above it. As the tube is sealed tightly, the HCN has nowhere to go and the evaporation continues for up to an hour until the air above the liquid is saturated with it. A small amount of this gas is then injected into the instrument and a detector such as the **nitrogen-phosphorus detector** (**NPD**) is used to identify the cyanide. If **GC-MS** is used, an extra step called **derivatisation** can be added in.[80] This involves attaching another chemical to the drug or poison to make a new version (a "derivative") that is more stable at the high temperatures inside the GC. It can also help toxicologists distinguish between very closely related members of the same family of drugs (such as opiates), which normally break into the same size fragments in the mass spectrometer, just by adding a few extra atoms.

Once cyanide has been detected and measured, understanding what that concentration means is fraught with difficulties. This is because cyanide concentrations can both increase and decrease after death, particularly if blood samples taken at the autopsy are not chemically preserved, or are left at room temperature.[61] In some cases, misleadingly high concentrations of cyanide can be produced by microbes after death when the metabolite thiocyanate is turned back into cyanide.[13,81] In others, there is a long time delay or even an exhumation when new information comes to light.[82] A new alternative marker for cyanide was discovered recently, **2-aminothiazoline-4-carboxylic acid**, which is much more stable and can be detected by **LC-MS**.[83]

Poor stability is just one of the reasons forensic toxicologists, despite what you might see on TV, don't try to estimate the dose a living person took of a drug or poison from a post-mortem blood concentration.[44] Stability, along with tolerance (Chapter 2), post-mortem redistribution (Chapter 4), collection tube problems (Chapter 5), drug interactions (Chapters 3 and 6), the state of the victim's health (Chapter 7), the presence of some naturally occurring drugs and poisons in the body (such as carbon monoxide, Chapter 12) and poison accumulation (Chapter 8) combine to make these calculations basically meaningless.

## 10.7 CASE CLOSED

We do not see the death of Henry Watkins in our case study, but we know from the dialogue it takes only minutes and that it involved a poison. As we have seen, there were many poisonous chemicals in dark-rooms that could have dispatched Mr Watkins, but the most likely is KCN, a fixer. We are never told what form the poison is in (liquid or solid) only that it is in a bottle. A solid is more likely, as these tended to have a longer shelf-life. Salts of cyanides dissolve easily in aqueous solutions[29] (such as drinks), and can indeed cause death in a matter of minutes, and also leave a few clues for the Pathologist; even if toxicology testing isn't carried out there may be a smell of bitter almonds and pink skin.

Adding the cyanide to sloe gin, apart from disguising the taste, is a nice poetic touch to the plot; not only do the pips of black-thorn fruits also contain cyanogenic alkaloids,[36] but in folk history the plant symbolises blessing to come after a challenge,[36] which was certainly the case for Henry Watkins' family.

## REFERENCES

1. P. Bottome, *The Liqueur Glass*, The Smart Set Publishing Co., London, 1915.
2. J. Emsley, *The Elements of Murder: A History of Poison*, Oxford University Press, Oxford, 2005.
3. U. Dassanayake and C. A. Gnanathasan, *J. Occup. Med. Toxicol.*, 2012, 7, 17.
4. F. Inkwright, *Botanical Curses and Poisons*, Liminal 11, London, 2021.
5. M. R. Cooper, A. W. Johnson and E. A. Dauncey, *Poisonous Plants and Fungi: An Illustrated Guide*, The Stationery Office, 2nd edn, Norwich, 2003.
6. D. Frohne and H. J. Pfänder, *Poisonous Plants: A Handbook for Doctors, Pharmacists, Toxicologists, Biologists and Veterinarians*, Manson Publishing, 2nd edn, London, 2004.
7. E. Walker, *Black-and-White Photographic Chemistry: A Reference*, NASA Technical Memorandum 87296, NASA, Cleveland, OH, 1986.
8. K. Harkup, *A is for Arsenic: The Poisons of Agatha Christie*, Bloomsbury, London, 2015.
9. R. Christison, *A Treatise on Poisons in Relation to Medical Jurisprudence, Physiology, and the Practice of Physic*, Barrington & Haswell, Philadelphia, PA, 1845.
10. A. A. Pappas, N. A. Massoll and D. J. Cannon, *Ann. Clin. Lab. Sci.*, 1999, **29**, 253–262.

11. J. F. O'Brien, *The Scientific Sherlock Holmes: Cracking the Case with Science & Forensics*, Oxford University Press, New York, 2013.
12. T. Hargreaves, *Poisons and Poisonings: Death by Stealth*, RSC Publishing, Cambridge, 2017.
13. *Disposition of Toxic Drugs and Chemicals in Man*, ed. R. C. Baselt, Biomedical Publications, Seal Beach, 12th edn, 2020.
14. M. Levine, A.-M. Ruha, K. Graeme, D. E. Brooks, J. Canning and S. C. Curry, *Chest*, 2011, **140**, 1357–1370.
15. L. Nelson, *J. Emerg. Nurs.*, 2006, **32**, S8–S11.
16. M.-H. Son, Y. Kim, Y.-H. Jo and M. Kwon, *Forensic Sci. Int.*, 2021, **328**, 111011.
17. D. Blum, *The Poisoner's Handbook: Murder and the Birth of Forensic Medicine in Jazz Age New York*, Penguin Books, London, 2010.
18. C. Valentine, *Murder Isn't Easy: The Forensics of Agatha Christie*, Sphere, London, 2021.
19. T. Stone and G. Darlington, *Pills, Potions, Poisons: How Drugs Work*, Oxford University Press, Oxford, 2000.
20. J. H. Bock and D. O. Norris, in *Forensic Plant Science*, ed. J. H. Bock and D. O. Norris, Academic Press, San Diego, 2016, ch. 1, pp. 1–22.
21. B. Hubbard, *Poison: The History of Potions, Powders and Murderous Practitioners*, Welbeck, London, 2020.
22. N. Panigrahi, S. P. Haranath, A. Ma, Y. Srinivas, S. Sirga, Ramkumar and K. Sarala, *Int. J. Crit. Care Med.*, 2019, **23**, 155–156.
23. L. S. Nelson, M. A. Howland, N. A. Lewin, S. W. Smith, L. R. Goldfrank and R. S. Hoffman, *Goldfrank's Toxicologic Emergencies*, McGraw Hill, New York, 11th edn, 2018.
24. J. P. André, *J. Chem. Educ.*, 2013, **90**, 352–357.
25. T. B. M. Hickey, J. A. MacNeil, C. Hansmeyer and M. J. Pickup, *Forensic Sci. Int.*, 2021, **326**, 110907.
26. D. E. Brooks, M. Levine, A. D. O'Connor, R. N. E. French and S. C. Curry, *Chest*, 2011, **140**, 1072–1085.
27. L. R. Lampariello, A. Cortelazzo, R. Guerranti, C. Sticozzi and G. Valacchi, *J. Tradit. Complementary Med.*, 2012, **2**, 331–339.
28. M. Grieve, *A Modern Herbal*, Tiger Books International, Twickenham, 3rd edn, 1998.
29. Medical Economics Company, *PDR for Herbal Medicines*, Medical Economics Company, Montvale, NJ, 1998.

30. M. Brown, *Death in the Garden: Poisonous Plants & Their Use Throughout History*, Pen & Sword Books Ltd, Barnsley, 2018.
31. R. Bevan-Jones, *Poisonous Plants: A Cultural and Social History*, Windgather Press, Oxford, 2009.
32. G. Zhang, H. Li, L. Sun, Y. Liu, Y. Cao, X. Ren and Y. Liu, *J. Chromatogr. Sci.*, 2023, **63**, 110–118.
33. W.-C. Chen, C.-H. Lee and H.-Y. Chen, *40th International Congress of the European Association of Poisons Centres and Clinical Toxicologists (EAPCCT)*, Tallinn, Estonia, 2020, p. 537.
34. P. M. North, *Poisonous Plants and Fungi*, Blandford Press, London, 1967.
35. H. Sharma and A. D. Meredith, *Emerg. Med. J.*, 2004, **21**, 392.
36. S. T. Dietz, *The Complete Language of Flowers: A Definitive and Illustrated History*, Wellfleet Press, New York, 2020.
37. F. Squire, *Foraging for Wild Foods*, IMM Lifestyle Books, Grantham, 2016.
38. P. Charlier, D. Corde, V. Bourdin, T. Martin, V. Tessier, M. Donnelly, A. Knapp and J.-C. Alvarez, *Forensic Sci., Med., Pathol.*, 2022, 244–250.
39. R. Burks, *Chem. World*, 2019, **16**, 71.
40. K. Harkup, *Chem. World*, 2020, **17**, 54–57.
41. W. Shakespeare, *Hamlet*, Oxford University Press, Oxford, 1998.
42. K. Harkup, *Death by Shakespeare: Snakebites, Stabbings and Broken Hearts*, Bloomsbury, London, 2020.
43. W. Shakespeare, in *Comedies*, David Campbell Publishers Ltd, London, 2000, ch. 5, vol. 1, pp. 408–511.
44. J. H. Trestrail III, *Criminal Poisoning: Investigational Guide for Law Enforcement, Toxicologists, Forensic Scientists and Attorneys*, Humana Press, Totowa, 2000.
45. J. Graham and J. Traylor, *Cyanide Toxicity*, StatPearls Publishing LLC, Treasure Island, FL, 2022.
46. M. Willes, *A Shakespearean Botanical*, Bodleian Library, Oxford, 2020.
47. C. Dickens, *The Life and Adventures of Martin Chuzzlewit*, Chapman & Hall, London, 1843.
48. J. K. Rowling, *Harry Potter and the Chamber of Secrets*, Bloomsbury, London, 1998.
49. D. J. Ballhorn, in *Nuts and Seeds in Health and Disease Prevention*, ed. V. R. Preedy, R. R. Watson and V. B. Patel, Academic Press, San Diego, CA, 2011, ch. 14, pp. 129–136.

50. E. F. Benson, *The Rubicon*, Methuen & Co., London, 1894.
51. A. Moore and D. Lloyd, *V for Vendetta*, DC Comics, Burbank, CA, 2005.
52. K. Takami, *Battle Royale*, Haika Soru, San Francisco, CA, 2nd edn, 2009.
53. S. Milazzo, S. Lejeune and E. Ernst, *Support. Care Cancer*, 2007, **15**, 583–595.
54. M. Moss, N. Khalil and J. Gray, *Can. Med. Assoc. J.*, 1981, **125**, 1126–1127.
55. A. H. Hall, C. H. Linden, K. W. Kulig and B. H. Rumack, *Pediatrics*, 1986, **78**, 269–272.
56. L. Sadoff, K. Fuchs and J. Hollander, *J. Am. Med. Assoc.*, 1978, **239**, 1532.
57. J. Avella, M. Katz and M. Lehrer, *J. Anal. Toxicol.*, 2007, **31**, 540–542.
58. M. J. Yeoh and G. Braitberg, *J. Toxicol., Clin. Toxicol.*, 2004, **42**, 855–863.
59. K. Stamyr, G. Thelander, L. Ernstgård, J. Ahlner and G. Johanson, *Inhalation Toxicol.*, 2012, **24**, 194–199.
60. T. Grabowska, R. Skowronek, J. Nowicka and H. Sybirska, *Clin. Toxicol.*, 2012, **50**, 759–763.
61. *Clarke's Analysis of Drugs and Poisons*, ed. A. C. Moffatt, D. Osselton and B. Widdop, Pharmaceutical Press, London, 4th edn, 2011.
62. M. Schulz, A. Schmoldt, H. Andresen-Streichert and S. Iwersen-Bergmann, *Crit. Care*, 2020, **24**, 195.
63. S. Kennedy and K. C. Cahill, *Clin. Case Rep.*, 2020, **8**, 3566–3567.
64. J. Jones, M. J. McMullen and J. Dougherty, *Am. J. Emerg. Med.*, 1987, **5**, 317–321.
65. F. Raška, B. Lipový, M. Hladík and J. Holoubek, *Acta Chir. Plast.*, 2021, **63**, 185–189.
66. M. J. Koschel, *Am. J. Nurs.*, 2002, **102**, 39–42.
67. R. J. Dinis-Oliveira, F. Carvalho, J. A. Duarte, F. Remião, A. Marques, A. Santos and T. Magalhães, *Toxicol. Mech. Methods*, 2010, **20**, 363–414.
68. B. M. Singh, N. Coles, P. Lewis, R. A. Braithwaite, M. Nattrass and M. G. FitzGerald, *Postgrad. Med. J.*, 1989, **65**, 923–925.
69. I. Prieto, I. Pujol, C. Santiuste, R. Poyo-Guerrero and A. Diego, *Emerg. Med. J.*, 2005, **22**, 389–390.

70. F. S. Pelição, D. M. De Paula, É. D. Botelho, M. D. Peres, G. Hampel, J. F. Pissinate, J. C. L. Ambrósio and B. S. De Martinis, *J. Toxicol. Anal.*, 2018, **1**, 5.

71. D. Blok, L. Ambrose, L. Ouellette, E. Seif, B. Riley, B. Judge, A. Ziegler and J. Jones, *Am. J. Emerg. Med.*, 2020, **38**, 846–848.

72. O. Wachełko, A. Chłopaś-Konowałek, M. Zawadzki and P. Szpot, *J. Anal. Toxicol.*, 2022, **46**, e52–e59.

73. A. Akintonwa and O. L. Tunwashe, *Hum. Exp. Toxicol.*, 1992, **11**, 47–49.

74. M. Nnoli, L. Nwidu, P. A. Nwafor and I. Chukwuonye, *Iran. J. Toxicol.*, 2013, **7**, 831–835.

75. J. Emsley, *Molecules of Murder: Criminal Molecules and Classic Cases*, RSC Publishing, Cambridge, 2008.

76. J. C.-C. Yu and A. Mozayani, in *Toxicology of Cyanides and Cyanogens*, ed. A. H. Hall, G. E. Isom and G. A. Rockwood, John Wiley & Sons, New York, 2015, ch. 20, pp. 276–282.

77. A. M. Calafat and S. B. Stanfill, *J. Chromatogr. B: Biomed. Sci. Appl.*, 2002, **772**, 131–137.

78. M. Odoul, B. Fouillet, B. Nouri, R. Chambon and P. Chambon, *J. Anal. Toxicol.*, 1994, **18**, 205–207.

79. A. Yamaguchi and H. Miyaguchi, *J. Chromatogr. Sci.*, 2021, **59**, 1–6.

80. J. L. McAllister, R. J. Roby, B. Levine and D. Purser, *J. Anal. Toxicol.*, 2008, **32**, 612–620.

81. P. Zuccarello, G. Carnazza, C. Raffino and N. Barbera, *J. Forensic Sci.*, 2022, **67**, 1617–1623.

82. P. D. Maskell, *J. Forensic Sci.*, 2021, **66**, 1862–1870.

83. J. Giebułtowicz, M. Rużycka, M. Fudalej, P. Krajewski and P. Wroczyński, *Talanta*, 2016, **150**, 586–592.

# CHAPTER 11

# Mad as a Hatter

If you see a term that's **bold** it's defined in the Glossary. Only the first time that the word appears in the chapter will it be indicated in this way.

---

**Case History: Workplace Poisoning**

*A young girl named Alice has fallen through a rabbit hole, drunk a liquid from a bottle to make her shrink, eaten a cake to make her grow, met various talking animals, and is now at a tea party with a Hatter, a dormouse and a hare.*
The Hatter was the first to break the silence. "What day of the month is it?" he said, turning to Alice: he had taken his watch out of his pocket, and was looking at it uneasily, shaking it every now and then, and holding it to his ear. Alice considered a little, and then said "The fourth."
"Two days wrong!" sighed the Hatter. "I told you butter wouldn't suit the works!" he added looking angrily at the March Hare.
"It was the *best* butter," the March Hare meekly replied.
"Yes, but some crumbs must have got in as well," the Hatter grumbled: "you shouldn't have put it in with the bread-knife."...
Alice had been looking over his shoulder with some curiosity. "What a funny watch!" she remarked. "It tells the day of the month, and doesn't tell what o'clock it is!" "Why should it?" muttered the Hatter. "Does *your* watch tell you what year it is?"

Chapter 7, A Mad Tea-Party

*Alice's Adventures in Wonderland* by Lewis Carroll,[1] 1865

---

Poisonous Tales: A Forensic Examination of Poisons in Fiction
By Hilary Hamnett
© Hilary Hamnett 2023
Published by the Royal Society of Chemistry, www.rsc.org

## 11.1   THE INVESTIGATION

The exchange between Alice and the Hatter (Figure 11.1) in the scene makes very little sense to her or to us as the reader, but it is thought that Lewis Carroll did not invent a mad character who happened to be a hatter (also known as a "milliner"), he was showing a type of behaviour that was common among hat makers in the 19th Century, leading them to be labelled 'mad'[2]

Alice...tried another question. "What sort of people live about here?"
"In that direction," the Cat said, waving its right paw round, "lives a Hatter: and in that direction," waving the other paw, "lives a March Hare. Visit either you like: they're both mad."
"But I don't want to go among mad people," Alice remarked.
"Oh, you can't help that," said the Cat: "we're all mad here. I'm mad. You're mad."
"How do you know I'm mad?" said Alice.
"You must be," said the Cat, "or you wouldn't have come here."

*Alice's Adventures in Wonderland* by Lewis Carroll,[1] 1865

So what is the cause of the Hatter's madness? We now know that it was mercury. Hatters were exposed to mercury compounds

**Figure 11.1**   Mad Hatter and the Rabbit, by John Tenniel.

as liquids and dust when they were making felt. In the 1800s this would have been made from the real fur of small animals such as rabbits, hares and beavers. The skin was soaked in or brushed with an acidic solution of mercuric nitrate [$Hg(NO_3)_2$] to separate it from the fur, which was then matted together into felt. During drying, mercury vapour would have been released into the air and inhaled by the craftsman. Once dry, the hatters then handled the fur multiple times, all the while breathing in the dust. Over months and years, this gave rise to symptoms such as erratic behaviour, instability, paranoia and irritability, which together were known as "erethism" or Mad Hatter's Disease. Of course, it was a while before we connected these problems with the mercury they were using day in, day out.[3,4]

This is an example of "chronic" poisoning. We have met the terms "chronic" and "acute" in this book already, usually to describe diseases or symptoms, but they can also be used to describe poisonings. All of the case studies we have looked at in previous chapters involved a person taking a large dose of a drug or poison and suffering serious effects quickly (within minutes or hours). These were "acute" poisonings. When it comes to metals such as mercury and arsenic, whilst acute poisoning is definitely possible, it's more common for people to take in small doses for months or even years, and suffer less severe symptoms but for longer; this pattern of poisoning is described as "chronic".

Today, if someone suffers from chronic mercury poisoning in the workplace, a forensic toxicologist is unlikely to be involved. Typically, the worker would go to their GP or occupational health team, who could order medical tests. Like many poisons, mercury disappears from the blood quickly, so a urine sample collected over the course of 24 h is a better way to spot long-term poisoning with mercury (it can stay in the urine for 6–12 months[5]). If someone is exposed to a large dose of mercury in the workplace because of an accident or a leak for example, they would be seen in hospital by a clinical toxicologist. That doesn't mean forensic toxicologists stay out of workplaces altogether – there is a big industry in workplace drug testing where potential or existing employees are checked for recreational drug use. Practices and employment laws vary between countries, but in the UK this is usually for safety-critical roles on public transport, construction sites and power lines. Testing can also be carried out after a workplace accident to check if anyone involved was impaired at the time. The tests are done on urine samples using "point-of-care" or dipstick devices based on **immunoassay** (see Chapter 6) in the workplace itself. The result is a simple yes or no, given by a colour change, and usually only looks for a limited number of the most common drug families (known as a "panel"). If a worker tests presumptively positive (see Chapter 6), their drug use needs to be confirmed by a laboratory test, and this is where forensic toxicologists can get involved. There are many ingenious ways workers try to cheat urine drug tests (to give a negative result even if they have used drugs recently). These include simply filling the testing cup with water from the sink or toilet, using someone else's urine, or adding 'cleansing' chemicals to their sample. These strategies can be easily picked up during testing by supervising the person giving the sample, or testing the colour, temperature or pH of the urine, disabling the taps in the collection room and adding coloured chemicals to the toilet cistern. People have therefore turned to the many websites that will sell you 'detox' chemicals to drink before giving the urine sample; these are unlikely to work however, and could be hazardous to the drinker.

Interestingly, there are websites selling quick urine tests for people worried about their mercury levels, which are also based on a colour change. As we will see, forensic toxicologists still encounter mercury in fatal poisonings, which can be acute or chronic, accidental or deliberate.

### 11.2   THE POISON BEHIND THE STORY

The mercury compound that poisoned the hatter in *Alice in Wonderland* was an inorganic salt, but you are probably more familiar with the mercury metal found in older thermometers and barometers. This is the shiny, silvery metallic liquid, which looks harmless, but if left out in the open it slowly evaporates and is breathed in.[2] You may also have a mercury amalgam filling in your teeth, where it has been mixed with silver, tin and copper. Debate still rages about their safety, and they are now restricted to adults who aren't pregnant or breastfeeding, with plans to gradually phase them out in the UK.

Mercury, like other metals, is mined from rocks containing small amounts of it, known as "ore". The most common ore is the scarlet mercury sulfide (HgS), also known as the pigment "cinnabar" (Figure 11.2), which is heated to evaporate the mercury.[6] Cinnabar occasionally still turns up in tattoo dyes and inks (despite being banned) leading to mercury poisoning in those who have been tattooed. Dyes may also be contaminated with unsafe levels of metallic mercury.[7]

Although we now know how poisonous mercury is, small amounts of it can still be found everywhere; in food, water, energy-saving lightbulbs and button batteries.[2,4] We all have low levels of mercury in our bodies from these everyday exposures, but these do not appear to be harmful.

It was a different story for the hatters, who worked with hot mercury in confined and poorly ventilated spaces, with no personal protective equipment. The problems for these workers were exacerbated by a poor understanding of hand and dental hygiene, leading to them also accidentally ingesting mercury. Even when they started experiencing symptoms, it was usually not serious enough for an employee to stop work, or they were misdiagnosed as having an obscure nervous disease instead.[8] In any case, raising health & safety concerns with the hat factory owner was a risky business; employees had few rights in the 19th

**Figure 11.2** The red pigment cinnabar (mercury sulfide).

Century and were unlikely to be able to prove their symptoms were related to the mercury they were handling. It wasn't until the 1890s that deaths due to mercury in factories even had to be reported to the authorities. This is in stark contrast to the death of a chemistry professor in the workplace in the USA in 1997, after she was exposed to a mercury compound through a latex glove. Her death was investigated, and the college was fined thousands of dollars for their poor safety standards.

In that instance, the mercury compound was being used for a nuclear magnetic resonance (NMR) experiment, but they have had many uses over the years. For instance, mercuric chloride ($HgCl_2$) also known as "bichloride of mercury" and "corrosive sublimate" was used as a developer in photography (see Chapter 10), in skin lightening (or antifreckle) creams,[9] as a catalyst, a disinfectant, and a pesticide.[10] Liquid mercury is still used in the electrolysis of salt solution (NaCl) to make metallic sodium (called the "chlor-alkali" process) and lipsticks have been made from cinnabar for centuries.[3] A very fine grey powder, which was a mix of mercury and chalk, was used up to the 1940s to visualise hidden fingerprints at crime scenes, leading to poisoning among those detectives and scenes of crime officers who were breathing in the dust.[4]

How toxic mercury is depends on whether it is in the salt, metal or organic form. To make matters more confusing, they can convert between the different types inside the body.[11] Liquid mercury is not particularly poisonous; if you swallow it, it generally passes straight through the gut without being absorbed.[2] But if you are near a pool of spilt mercury, the vapour it releases absorbs very easily through the lungs and is then converted into $Hg^{2+}$ ions in the bloodstream. Mercury is made more poisonous by combining it with carbon to make organic mercury compounds such as dimethylmercury [$Hg(CH_3)_2$] or methyl mercury ($HgCH_3$). These are very easily absorbed through the gut, and once they have reached the brain can be turned back into liquid mercury. Mercury salts such as $HgCl_2$, on the other hand, are hard to dissolve in water and so are poorly absorbed and less poisonous, but can still damage the kidneys.[11]

All forms of mercury act in a similar way; they attach themselves to proteins containing **cysteine** molecules by replacing the –HS with –HgS.[10] This alters the structure of the protein, preventing it from doing its job properly. As these proteins are found all over the body, mercury quickly builds up in all the tissues, particularly the brain. When a lot of mercury vapour is inhaled all at once, bronchitis can lead to respiratory failure and death. If a large dose of a mercury salt is ingested, gastrointestinal symptoms such as nausea, abdominal pain, vomiting and diarrhoea are more likely.[11] But mercury is a sinister poison; some of its effects are delayed for weeks or even months after exposure to a large dose, especially weight loss and excessive urinating due to kidney damage.[3] With chronic mercury poisoning, symptoms such as mild tremors (known as the "hatters' shakes"), memory loss, anxiety, depression and paranoid delusions ("mercury madness"), blurred vision, and seizures have been reported.[10,12] It can also cause long-term inflammation of the mouth and gums (which can become covered in a grey film), drooling and loss of teeth.[2,3]

There are antidotes to mercury poisoning, known as "chelating agents" and they take advantage of how much mercury likes to bind to sulfur. The first one to be discovered was **dimercaprol**, which is injected into a muscle of the poisoning victim. Mercury binds to the S atoms and is carried out of the organs and the body through the urine.[3] Dimercaprol was originally designed to be used against arsenic, but has since been found to work well against mercury and other metals such as thallium. Other antidotes with similar structures have also been developed.[13]

Mercury has a long history as a medicine, particularly for what were known as "venereal" diseases, but are now called sexually transmitted infections (STIs). The name venereal comes from the Roman goddess of love, Venus, so it may have seemed romantic that such diseases were 'cured' by something associated with another Roman god, Mercury the messenger. In reality, these cures often did more harm than good. Mercurous chloride ($Hg_2Cl_2$) known as "calomel" or "sweet mercury" was widely used to treat the sores common in syphilis,[14] and to 'flush' the illness out through its effect on digestion.[15] There was also a treatment called "blue mass", one-third of which was mercury metal,[16] and a kind of plaster to apply to the sores containing mercury mixed into oil.[4] The "powdering tub of infamy" was, in fact, a tub or tent in which the patient was immersed in the fumes of cinnabar, which was burnt on a hot plate or hot coals. As the cinnabar evaporated it condensed and settled as a powder onto the patient's body.[17] Apart from the somewhat blasé approach of covering whole body with the cure (rather than just the sores), this method also carried a risk of carbon monoxide poisoning (see Chapter 12).

Mercury cures were replaced with arsenic in 1905,[18] but we now know syphilis to be a bacterial infection caused by *Treponema pallidum*, and today it would be treated with a course of antibiotics. Although germs were not understood back when these cures were popular, mercury was actually pretty effective against the bacteria; it bound to the S atoms in their proteins stopping them from reproducing, but only using it topically as a plaster meant the disease could not be cured completely. The side-effects were also very unpleasant.

Some mercury-based medicines could be bought "over-the-counter" until the mid-1950s when they were banned in the UK.[4] When forensic toxicologists talk about over-the-counter (OTC) drugs or medicines, they mean something you can go into a pharmacy and buy without needing a prescription. In some cases a staff member might ask you a few questions about your age, your symptoms, any other medications you are taking, *etc.* but nothing would go on your medical records. The most common OTC medicines seen in toxicology cases are pain relievers such as paracetamol, sleeping tablets, cough and cold medicines, and antihistamines. Before the 1955 ban, yellow mercuric oxide (HgO) was used in Golden Eye ointment for treating sties on the eyelid and conjunctivitis. Mercury was also the active ingredient in teething powders for babies (Steedman's Teething Powder was 26% calomel) producing the potentially fatal "pink disease" visible on the cheeks, fingers, toes, nose, and buttocks of a small number of babies.[4] In Chapter 4 we met another type of deadly teething powder containing belladonna.

## 11.3 POISONOUS PLOTS

Mercury as a poison is not a common choice in fiction, partly because its metallic taste makes it hard to disguise in food or drink. In fact, it doesn't even appear in a list of the top 73 poisons used in literature compiled by a medical toxicologist.[19]

It does make an appearance in the comic opera *Il Campanello di notte* (1836) ("The Night Bell") by Gaetano Donizetti, as a medicine. The plot involves a wealthy old pharmacist (Don Hannibal) marrying a beautiful young woman. Her jealous ex-fiancé Enrico tries to keep Don Hannibal out of the marital bed on their wedding night by calling at the pharmacy (disguised as a querulous old man called Henry) to ask him to make up a ridiculously complicated prescription for his fictional wife. He is relying on the legal requirement for pharmacists to answer their bell at any hour of the night.[20] The aria "La Povera Anastasia" lists all of the ingredients needed

HENRY:

Ma qui sta il re dei recipi

Che tutto guarirà

*[mostra la ricetta avvoltolata]*

Si prenda l'acquq celebre del gran monsù Maurizio con l'altra capo-cefalo e poi la fugia denica.

—

Con questa poi mischiatevi...

Mischiate, rimischiate, poi pillole formate.

HENRY:

Here is the king of prescriptions, which cures every ailment.

*[he reads from the recipe]*

You must take the celebrated Murizio water, then the fugrademica, mix it with...

*[seven more ingredients are listed]*

...then mix them together, and roll them into pills.

DON HAN:

Ma questi sono liquidi.

DON HAN:

But all these are in liquids.

HENRY:

Che ad una, a quattro, a sette

Si devono ingojar. —

HENRY:

And must be taken by four, by seven—

DON HAN:

Basta! —

DON HAN:

Enough—enough

| HENRY: | HENRY: |
|---|---|
| Poi l'ombelìo di Venere | Take afterwards, antimonial butter... |
| Burirro d'antimonia, | *[even more ingredients]* |
| L'estratto di cicuta | extract of chicory and poppies, |
| Papeveri, la ruta | the æthio mineral, cordial syrup... |
| L'etiope minerale, | *[the list goes on]* |
| Siroppo cordiale. — | |

Scene xii
*Il Campanello de notte* by Gaetano Donizetti,[21] 1861

You may recognise the poppy extract from Chapter 2, but the ingredient that interests us is "L'etiope minerale", also known as HgS or cinnabar (Figure 11.2). As we will see in Section 11.5, cinnabar is still added to some herbal medications today. The combination of ingredients in Henry's made-up prescription would have been more of a hazard than a healer, so it's fortunate that Don Hannibal refuses to make it up in the end.[22] This idea of making up a medication from scratch might seem odd to those of us who have grown up seeing neat boxes of mass-produced pills in pharmacies, but during the time the opera was composed, concoctions were mixed together by individual pharmacists.[23]

Comic books have used the fascinating properties of metallic mercury to create superheroes and villains. DC Comics had a super-speedy fictional superhero called Max Mercury, as well has a supervillain called the Mad Hatter who was an enemy of Batman. The teenage girl Cessily Kincaid, known as Mercury, in Marvel's *X-Men* series is a human mutant with the power to mimic liquid metallic mercury. This allows her to reshape or solidify at will and cling to surfaces. Another Marvel character is called Quicksilver (Pietro Maximoff) and he can move at great speed. The *X-Men* character Wolverine's backstory involves a metal being grafted to his bones and claws. In an interesting example of how poisons in fiction can lead to real toxicology cases, two 15-year-olds (one male and one female) in Sri Lanka in 2016 and India in 2014, injected themselves in the arms with mercury after watching *X-Men Origins: Wolverine* (2009).[24-26] The male teenager already had a history of spider bites after watching *Spiderman*.

## 11.4  MERCURY AND MAGIC

Mercury metal is believed to have enormous spiritual and magical powers.[9] It was consumed as an elixir in Ancient India, China[27] and Egypt[2] and is still sold as *azogue* in religious stores, or botanicas, for use in some African–Caribbean spiritual rituals. It is typically found in capsules that contain 8–9 g of metallic mercury and are designed for carrying around in a sealed pouch or pocket as an amulet. Other spiritual practices include sprinkling mercury on the floor, swallowing it (sometimes mixed with egg, port, nutmeg and milk), boiling it in a pot, adding it to baths, burning it in a candle and mixing it with perfume. These practices are supposed to bring happiness, luck in love, money, or health and to ward off evil.[18,28,29] Mercury is even injected subcutaneously by the superstitious to bring good luck when travelling abroad.[30] This can have unpleasant side-effects, and in 2004 a couple from the Republic of Honduras went to hospital in the USA after being injected multiple times in the hands with mercury to ward off evil and protect against any unknown diseases when travelling.[29]

Unfortunately all this activity with mercury can lead to dangerously high levels of fumes in the home as it falls into cracks and slowly vapourises. This is not only harmful to current and future residents, but also to those in neighbouring apartments. Much of the mercury used in religious practice is also disposed of improperly, going down the drain for example.[28]

In Europe, alchemists tried unsuccessfully to convert mercury into gold by combining it with other metals.[31] There was also a widespread medieval belief that mercury protected against the evil eye, spells, demons and misfortunes.[32]

## 11.5  MODERN MEDICAL USES

We met Ayurvedic medicines, which originate in India, in previous chapters. These can be herbal-only medicines, such as *guggul* and *triphala*, or herbo-mineral or *rasa shastra* medicines where herbs are combined with metals such as lead (*naga*) or mercury (*parada*). The addition of metals to herbal preparations is attractive as it leads to lower doses being needed and long shelf-lives. Mercury is usually added as cinnabar.[33] For example, *Rasasindura* is made from cinnabar, honey, milk and butter.[10] Ayurveda practitioners believe that if metal-containing medicines are prepared according to Ancient protocols, the metals are detoxified or "defanged". The detoxification involves a complex oxidation process of heating and boiling.[34] However, a study in 2008 tested the amounts of metal in Ayurvedic medicines purchased on the internet. More than 20% of them contained concentrations of metals above Western acceptable daily intakes.[35]

There is also a theory that as the mercury is in its inorganic form, it is much less toxic than using mercury metal. Testing of some medicines in Canada in 2013, confirmed that the amount of mercury that could be absorbed into the bloodstream from these medicines was generally low.[36] Nevertheless, poisonings still happen; in 2020 in Canada, a 29-year-old male went to A&E with chronic back pain, weakness, numbness, and tingling. After 45 days of using an Ayurvedic medicine for his back pain, he had developed mercury toxicity and needed to be treated with dimercaprol (see Section 11.2).[37]

In Traditional Chinese Medicine (TCM), products can contain metallic mercury (known as *Shui Yin*).[38,39] Cinnabar (*Zhu Sha*) is also deliberately added to around 40 medicines (called "mercurials"), and these are used as antidotes for insect bites or as sedatives.[38,40]

The mercury metal inside thermometers has been replaced by red or blue dyed alcohol now, because of the risk of the thermometer breaking during use. Although we might think the risk is of young children swallowing the mercury, it is more likely to be a sharps injury to the skin. In China in 2012, a 1-year-old boy was injured under the arm while his temperature was being taken. The mercury was sucked out of him using a needle and he suffered no long-term effects.[41] In another case, a 9-year-old boy was injured in the hand by a broken mercury thermometer in Slovakia in 2018.[42] Although these thermometers are now banned from sale in the EU, we will see in Section 11.6, that there are still plenty lying around in bathroom cabinets.

The same mercuric nitrate that caused the hatters to become ill, was also once used as an early test for vitamin B12 deficiency, although this has now been replaced by **immunoassays** (see

Chapter 6). A controversial use of mercury was in thimerosal, which started being added to vaccines in the 1930s to preserve them,[4] but was phased out in the UK in 2004. The antiseptic merbromin, an organic mercury compound is sold as a 1–2% solution to apply to the skin. After several high-profile poisonings, one involving the death of a newborn baby, it was declared unsafe in the USA in 1998.[16] It is still sold in other countries and, of course, on the internet.

## 11.6   MODERN TOXICOLOGY CASES

Although we might think chronic workplace poisoning with mercury is a thing of the past, in 2012 in the USA, a 36-year-old man went to hospital with a rash, a cough and joint pains. He had been employed to recycle thermometers and the metallic mercury he had been breathing in had deposited in his heart. When he was admitted, his blood mercury level was 0.2 mg $L^{-1}$ (compared to a normal level of 0.002 mg $L^{-1}$[16]) and even though he was given a chelating agent, he died after 18 days.[43]

Thankfully these cases are now rare, but toxicologists still see mercury in suicides, homicides and accidents. Those taking mercury (knowingly or otherwise) usually inhale the vapour, eat a mercury salt, or, in recent years, there has been an increase in self-injection of metallic mercury. This can be for self-harm, or by athletes hoping to build muscle tone, or strangely enough as an aphrodisiac.[30]

In Greece in 2014, a 22-year-old female injected herself in the forearm with mercury taken from three broken thermometers. She mixed the metallic mercury with alcohol for injection. Once mercury is injected intravenously, it can quickly cross over into the systemic circulation and be distributed into the tissues, particularly the lungs. Scans of the patient showed mercury deposits in her lungs and kidneys. She was treated with a chelating agent and discharged, but continued to experience seizures.[44,45] A 61-year-old Guyanese man developed a tremor, and burning pain in his feet after injecting himself with mercury as a cure for his diabetes. He was treated with dimercaprol and recovered, but his tremor took nine months to disappear (and his diabetes remained).[46]

In Portugal in 2016, a 53-year-old female chemistry teacher was taken to hospital after having taken mercuric oxide (HgO) in a suicide attempt. She was confused, and had abdominal pain and vomiting. Her mercury blood concentration was 6.8 mg $L^{-1}$ (compared to a potentially fatal level of 0.5 mg $L^{-1}$[47]) and she died after 15 days.[48]

A 14-year-old girl in Poland in 2014 suddenly died during one of many visits to hospital. She had been suffering from stomach pains, vomiting and multi-organ failure after eating, inhaling and rubbing mercury metal into her skin (all from broken thermometers). After her death it was revealed that she was suffering from Münchausen syndrome, where people deliberately make themselves ill for attention.[9]

Of course, not everyone is aware they are breathing in mercury vapour, and a 19-month-old girl was taken to hospital in the USA in 2020 with a cough, fever and hypoxia. Her blood mercury level was 0.15 mg $L^{-1}$ and others living in her family's mobile home also had elevated mercury levels after it was found coming from a sink drain. She spent over 50 days in hospital.[49] None of the adults living in the same house were quite so badly affected, probably because children tend to build up higher levels of mercury, even when exposed to the same dose.[18]

Using skin creams containing significant amounts of mercury can lead to unpleasant side-effects and high levels of mercury in the urine (0.03 mg $L^{-1}$ compared to a normal level of 0.005 mg $L^{-1}$[5]).[16,40] These creams are desirable to people because of their bleaching or lightening effects, and are banned in many countries, but of course can be bought on the internet. Skin-lightening soaps containing mercury iodide ($Hg_2I_2$) are available in some African countries, but the WHO has warned against using them.[50] A gel for the relief of eczema containing mercury bromide ($Hg_2Br_2$) was applied by a 77-year-old female in Portugal in 2009 to her chest. She was taken to hospital and died after 7 days.[51]

There seems to be a long association of mercury and sexual performance,[32] and in India in 2005, a 20-year-old male was admitted to hospital after injecting himself with mercury in the hand as an aphrodisiac. Far from the desired effects he had been hoping for, he had a fever, racing heartbeat, and swelling of the injection site.[52] Some people are prepared to resort to more drastic injection sites, including a 72-year-old man in Korea in 2006 who was injected in the penis in an attempt at augmentation. He went to hospital complaining of pain, fever and nausea.

The damage and build-up of mercury was so severe that a complete amputation was needed.[53]

There is also the more sinister use of mercury as a murder weapon. In a strange case in Japan in 2019, a 36-year-old man went to hospital after finding a small silver object in a cigarette he had received from an acquaintance. He had already smoked 13 cigarettes from the same box and started to get a headache, lose his appetite, and feel fluey. His blood mercury was 0.1 mg L$^{-1}$, but he was eventually discharged. Police later discovered that it was an attempted murder.[54] In Canada in 2019, two cases where mercury was added to food in deliberate poisoning attempts were recorded. In one case it was added to a landlord's milk after a dispute with his boarder. In another, a man poisoned his wife and seven-year-old daughter by adding mercury to some rice pudding. Fortunately both suffered only nausea.[55]

In China in 2017 a 34-year-old male went to hospital after his girlfriend (a nurse) drugged his coffee with hypnotics and once he fell asleep injected him with 40 g of mercury (taken from broken thermometers) in his elbow. After four months of treatment he was finally discharged.[56]

In Chapter 8 we examined how toxicology can still be done on exhumed bodies, and how we can find poisons such as **strychnine**. In China, a 33-year-old woman was found dead in bed in 2012 by her husband (a doctor) after being unwell for months with a cough, chest pain and fatigue. Because of this history of illness, there was no autopsy or toxicology and she was buried. Two months after her burial, news of her husband's extra-marital affair reached her family and she was exhumed. The Pathologist found mercury metal in her lungs and blood along with **cyanide** (Chapter 10). The husband confessed that he had injected her with a mix of glucose solution and mercury. When that didn't work, he had moved onto cyanide. We will see in Chapter 12 that at autopsy, the tests conducted for drugs and poisons can be limited depending on the circumstances. In this case (and many others) there were no tests at all at the time of death. This might seem like an odd decision, but if a natural cause of death is identified as likely, sometimes expensive toxicology tests are deemed unnecessary.

In Poland in 2015, a 53-year-old female died with the cause of death being given as "heart attack". No toxicology tests were performed despite her being hospitalized twice prior to her death with poisoning symptoms (vomiting, diarrhoea and kidney failure). Five years after she was buried, police searched the workplace of her son-in-law and found a stash of poisons, antidotes and a handbook on toxicology. Her body was exhumed and unusually high levels of mercury were found in her hair.[57]

Hair testing is particularly useful in cases of exhumation as while body fluid samples dry up quickly after death, hair takes much longer to decompose. It can also be used to test live people, typically in Family Court cases where parents have to prove they have stopped using drugs in order to win back custody of their children. Hair and heavy metals have a long history together, and it was back in the 1960s that hair samples were first used to look for exposure to metals. As our hair grows, any drugs or poisons we are exposed to are incorporated into the shaft from the bloodstream or sweat, and are very hard to get rid of (although some people try with bleaching and so-called "detox" shampoos). Approximately 1 cm of head hair corresponds to 1 month of growth, so if a person has long hair we can look at a year or longer of exposure.[58] Hair is particularly useful for mercury testing as it has many –SH groups that Hg likes to bind to.[10]

A major drawback of hair testing for any drug or poison is the possibility of environmental contamination. Just because a drug or poison is detected in a hair sample, does not necessarily mean that it was actively consumed. Drugs and poisons that are gases or vapour can settle on the outside of the hair, and if the sample is not thoroughly washed before it is analysed, this can lead to a false-positive.

### 11.7   CATCHING THE CHEMICAL CULPRIT

Today we take toxicology tests for drugs and poisons in body fluid samples for granted, but they weren't invented until the mid-1800s. While scientists knew about many toxic alkaloids, they didn't know how to test for them in the body.[16] One of the early tests was called the Reinsch test (published in 1841) and could detect metal poisons including arsenic and mercury.[59] These were 'wet chemistry' techniques that involved dissolving the body tissue in acid then inserting a small piece of copper into the solution. Any metal present would 'plate out' or cling to the copper, giving it a new coating. In the case of mercury, it was a very distinctive silvery new coat.

Nowadays we used instrumental techniques such as **Atomic Absorption Spectroscopy (AAS)** to detect metals.[9,16] Many of the techniques we have looked at in previous chapters can be used for multiple different drugs, but AAS looks only for metals, and can only see a single element at a time.[5] That first step of adding acid to the sample and sometimes heating it is still needed. After that, the solution is passed into an "atomiser" to produce a very fine mist of tiny droplets (similar to perfume). The mist is mixed with a flammable gas and burnt in a flame.[3] A specific type of light is passed through the metal in the sample and a detector measures how much is absorbed.[59] AAS can tell us which metal was involved, but can't distinguish between the different forms of mercury we have learned about in this chapter.[15] To do that we need another technique, such as **Gas Chromatography (GC)**, see Chapter 3.

## 11.8 CASE CLOSED

So was the Mad Hatter's bizarre behaviour a result of mercury poisoning? It is certainly the case that many hatters developed neurological symptoms from chronic exposure to mercury. It was not just a problem for hatters – other people who worked with mercury every day, such as miners and dentists also suffered.[4] As did gilders (people who coated objects with gold) and jewellers, who both used mercury to dissolve small amounts of gold.[31]

As we have seen, hatters became unstable, paranoid and irritable after handling mercury for long periods. But other symptoms of chronic mercury poisoning were shyness,[12] timidity, and a desire to go unnoticed. These are not really consistent with the portrayal of Lewis Carroll's Mad Hatter, who is more of an eccentric extrovert. Another theory is that he was a caricature of an Oxford furniture dealer called Theophilus Carter, who was a clock inventor, an eccentric, and always wore a top hat.[60]

The problem of mercury poisoning in hat-making was gradually reduced by changes in fashion towards silk hats, and was very quickly resolved by the arrival of World War II, when mercury became so valuable for detonators that it was no longer used for hats.[61]

# REFERENCES

1. L. Carroll, *Alice's Adventures in Wonderland*, Macmillan, London, 1865.
2. B. Hubbard, *Poison: The History of Potions, Powders and Murderous Practitioners*, Welbeck, London, 2020.
3. T. Hargreaves, *Poisons and Poisonings: Death by Stealth*, RSC Publishing, Cambridge, 2017.
4. J. Emsley, *The Elements of Murder: A History of Poison*, Oxford University Press, Oxford, 2005.
5. *Clarke's Analysis of Drugs and Poisons*, ed. A. C. Moffatt, D. Osselton and B. Widdop, Pharmaceutical Press, London, 4th edn, 2011.
6. M. Marchini, M. Gandolfi, L. Maini, L. Raggetti and M. Martelli, *Proc. Natl. Acad. Sci. U. S. A.*, 2022, **119**, e2123171119.
7. A. Prantsidis, N. Raikos, I. Pantelakis, K. Spagou and E. Tsoukali, *Hippokratia*, 2017, **21**, 197–200.
8. F. E. Tylecote, *Lancet*, 1912, **180**, 1137–1140.
9. T. Lech, *Forensic Sci. Int.*, 2014, **237**, e1–e5.
10. S. Cappelletti, D. Piacentino, V. Fineschi, P. Frati, S. D'Errico and M. Aromatario, *Crit. Rev. Toxicol.*, 2019, **49**, 329–341.
11. R. A. Bernhoft, *J. Environ. Public Health*, 2012, **2012**, 460508.
12. J. Timbrell, *The Poison Paradox: Chemicals as Friends and Foes*, Oxford University Press, Oxford, 2005.
13. M. Rafati-Rahimzadeh, M. Rafati-Rahimzadeh, S. Kazemi and A. A. Moghadamnia, *Daru, J. Fac. Pharm., Tehran Univ. Med. Sci.*, 2014, **22**, 46.
14. K. Harkup, *Death by Shakespeare: Snakebites, Stabbings and Broken Hearts*, Bloomsbury, London, 2020.
15. L. S. Nelson, M. A. Howland, N. A. Lewin, S. W. Smith, L. R. Goldfrank and R. S. Hoffman, *Goldfrank's Toxicologic Emergencies*, McGraw Hill, New York, 11th edn, 2018.
16. *Disposition of Toxic Drugs and Chemicals in Man*, ed. R. C. Baselt, Biomedical Publications, Seal Beach, 12th edn, 2020.
17. A. C. Kail, *Med. J. Aust.*, 1983, **2**, 445–449.
18. P. O. Ozuah, *Curr. Probl. Pediatr.*, 2000, **30**, 91–99.
19. J. H. Trestrail III, *Criminal Poisoning: Investigational Guide for Law Enforcement, Toxicologists, Forensic Scientists and Attorneys*, Humana Press, Totowa, 2000.
20. *A New Grove Dictionary of Opera*, ed. S. Sadie, Oxford University Press, Oxford, 1997.

21. G. Donizetti, *Il Campanello: A Comic Operetta*, W. Jeffs, London, 1861.
22. J. P. André, *J. Chem. Educ.*, 2013, **90**, 352–357.
23. C. Valentine, *Murder Isn't Easy: The Forensics of Agatha Christie*, Sphere, London, 2021.
24. D. Sukheeja, P. Kumar, M. Singhal and A. Subramanian, *J. Lab. Physicians*, 2014, **6**, 55–57.
25. S. M. Thanuja Nilushi Priyangika, W. G. S. G. Karunarathna, I. Liyanage, M. Gunawardana, B. Dissanayake, S. Udumalgala, C. Rosa, T. Samarasinghe, P. Wijesinghe and A. Kulatunga, *BMC Res. Notes*, 2016, **9**, 189.
26. M. Wijesinghe, S. Gouse, G. Rajapakse, T. Beneragama and D. Perera, *Sri Lanka J. Surg.*, 2015, **33**, 25–26.
27. F. Campbell, in *Chemistry in its Element*, RSC Publishing, Cambridge, 2009.
28. D. M. Riley, C. A. Newby, T. O. Leal-Almeraz and V. M. Thomas, *Environ. Health Perspect.*, 2001, **109**, 779–784.
29. L. Prasad Venkat, *Environ. Health Perspect.*, 2004, **112**, 1326–1328.
30. U. Da Broi, C. Moreschi, A. Colatutto, B. Marcon and S. Zago, *J. Forensic Leg. Med.*, 2017, **50**, 12–19.
31. J. D. Blum, *Nat. Chem.*, 2013, **5**, 1066.
32. C. R. Rider, PhD thesis, University College London, 2016.
33. A. Kadam, *Rasamruta*, 2013, **2013**, 1–7.
34. A. Chopra and V. V. Doiphode, *Med. Clin. North Am.*, 2002, **86**, 75–89.
35. R. B. Saper, R. S. Phillips, A. Sehgal, N. Khouri, R. B. Davis, J. Paquin, V. Thuppil and S. N. Kales, *J. Am. Med. Assoc.*, 2008, **300**, 915–923.
36. I. Koch, M. Moriarty, J. Sui, A. Rutter, R. B. Saper and K. J. Reimer, *Sci. Total Environ.*, 2013, **454–455**, 9–15.
37. F. Saleh, D. Ovakim, M. Yarema, M. Riggan, R. Hartmann, S. Lucyk and L. Lum, *American College of Medical Toxicology 2020 Annual Scientific Meeting New York, NY*, 2020, p. 164.
38. J. Liu, J.-Z. Shi, L.-M. Yu, R. A. Goyer and M. P. Waalkes, *Exp. Biol. Med.*, 2008, **233**, 810–817.
39. E. Ernst, *Pharmacoepidemiol. Drug Saf.*, 2004, **13**, 767–771.
40. L. E. Davis, *West J. Med.*, 2000, **173**, 19.
41. Q. H. Zhu, Y. Chen, Q. L. Zeng and J. B. Zhao, *Clin. Radiol.*, 2012, **67**, 83–85.

42. S. Plačková, J. Kresanek, B. Caganova, O. Otrubova, I. Batora, P. Banovcin, P. Durdik and M. Molnar, *40th International Congress of the European Association of Poisons Centres and Clinical Toxicologists (EAPCCT)*, Tallinn, Estonia, 2020, p. 582.

43. T. Alhamad, J. Rooney, A. Nwosu, J. MacCombs, Y.-s. Kim and V. Shukla, *Int. J. Urol. Nephrol.*, 2012, **44**, 647–651.

44. K. Fragkou, C. Marvaki, G. Papantoniou, E. Zisiou and E. Triantafyllou, *Health Sci. J.*, 2014, **8**, 541–547.

45. S. Karatapanis, F. Lamprianou, G. Ntetskas and A. Kotis, *Br. Med. J. Case Rep.*, 2015, **2015**, bcr2014207075.

46. H. H. Schaumburg, C. Gellido, S. W. Smith, L. S. Nelson and R. S. Hoffman, *Neurology*, 2009, **72**, 377–378.

47. M. Schulz, A. Schmoldt, H. Andresen-Streichert and S. Iwersen-Bergmann, *Crit. Care*, 2020, **24**, 195.

48. D. Dias, J. Bessa, S. Guimarães, M. E. Soares, M. d. L. Bastos and H. M. Teixeira, *Forensic Sci. Int.*, 2016, **259**, e20–e24.

49. W. J. Meggs, T. Byerly, K. Ziemba, K. Shum, M. Dexter, M. Ledoux, R. Langley, A. R. Dulaney and M. Beuhler, *American College of Medical Toxicology 2020 Annual Scientific Meeting New York, NY*, 2020, p. 125.

50. J. Emsley, *Vanity, Vitality, and Virility: The Science Behind the Products you Love to Buy*, Oxford University Press, Oxford, 2004.

51. P. Triunfante, M. E. Soares, A. Santos, S. Tavares, H. Carmo and M. de Lourdes Bastos, *Forensic Sci. Int.*, 2009, **184**, e1–e6.

52. A. Gopalakrishna and T. V. Pavan Kumar, *Indian J. Plast. Surg.*, 2008, **41**, 214–218.

53. K. J. Oh, K. Park, T. W. Kang, D. D. Kwon and S. B. Ryu, *Urology*, 2007, **69**, 185.e183–185.e184.

54. M. Hitosugi, M. Tojo, M. Kane, N. Shiomi, T. Shimizu and T. Nomiyama, *Int. J. Leg. Med.*, 2019, **133**, 479–481.

55. K. Kenny, M. Sandercock and J. Webster, *Can. Soc. Forensic Sci. J.*, 2019, **52**, 122–128.

56. Q. Lu, Z. Liu and X. Chen, *Medicine*, 2017, **96**, 46.

57. T. Lech, *Am. J. Forensic Med. Pathol.*, 2015, **36**, 227–231.

58. E. Gallardo and J. A. Queiroz, *Biomed. Chromatogr.*, 2008, **22**, 795–821.

59. *More Chemistry and Crime: From Marsh Arsenic Test to DNA Profile*, ed. S. M. Gerber and R. Safterstein, American Chemical Society, Washington DC, 1997.

60. H. A. Waldron, *Br. Med. J.*, 1983, **287**, 1961.

61. R. P. Wedeen, *Am. J. Indian Med.*, 1989, **16**, 225–233.

# CHAPTER 12

# The Silent Killer

If you see a term that's **bold** it's defined in the Glossary. Only the first time that the word appears in the chapter will it be indicated in this way.

---

**Case History: Domestic Poisoning**

*The narrator receives a letter from a childhood friend Roderick Usher complaining of an illness and asking him to visit the family home, called the House of Usher. When the narrator arrives, he sees the house in disrepair and listens to Roderick describe his symptoms.*

He entered, at some length, into what he conceived to be the nature of his malady. It was, he said, a constitutional and a family evil, and one for which he despaired to find a remedy—a mere nervous affection, he immediately added, which would undoubtedly soon pass off. It displayed itself in a host of unnatural sensations. Some of these, as he detailed them, interested and bewildered me...He suffered much from a morbid acuteness of the senses; the most insipid food was alone endurable; he could wear only garments of certain texture; the odors of all flowers were oppressive; his eyes were tortured by even a faint light

*The Fall of the House of Usher* by Edgar Allan Poe,[1] 1839

---

Poisonous Tales: A Forensic Examination of Poisons in Fiction
By Hilary Hamnett
© Hilary Hamnett 2023
Published by the Royal Society of Chemistry, www.rsc.org

## 12.1  THE INVESTIGATION

This list of symptoms appears a bit vague at first glance, and Roderick himself tries to pass off the illness as an inherited nervous affection, but viewed through the eyes of a toxicologist, it looks like a kind of slow poisoning. Roderick could be having his food or drink poisoned, but the important clue here is that someone else in the house is also unwell, his sister Madeline

The disease of the lady Madeline had long baffled the skill of her physicians. A settled apathy, a gradual wasting away of the person, and frequent although transient affections of a partially cataleptical character, were the unusual diagnosis

*The Fall of the House of Usher* by Edgar Allan Poe,[1] 1839

If more than one person (or pet) in the house has symptoms it suggests the poison is in the environment, such as in the air or water. Toxicologists sometimes encounter cases where fumes from paint, carpet glue, crop spraying or toxic gases cause illness. The Ushers and the narrator (Figure 12.1) do not report an obvious smell, suggesting something stealthier is to blame.

When this short story was written (in 1839) heat in houses was supplied by coal fireplaces, and indoor lighting came from burning coal gas in lamps. Coal gas was a mix of hydrogen, methane and the very poisonous carbon monoxide (CO). We saw the difference between chronic and acute poisoning in Chapter 11, and Roderick and Madeline's symptoms look like a case of chronic CO poisoning. In other parts of the story Roderick is described as having a "ghastly pallor" or being "cadaverously wan", as walking with an "unequal step", having an "agitated mind" and appearing "emaciated", which as we will see are all symptoms of chronic CO poisoning.

You have probably been warned about the dangers of acute CO poisoning. Safety posters, TV adverts and campaigns give us a list of symptoms to watch out for: headaches, dizziness, breathlessness, coma and eventually death. But we know much less about low-level chronic CO poisoning, despite the fact that many people in the 19th Century probably suffered from it because of leaking gaslights and blocked chimneys. From case studies where chronic CO poisoning *has* been diagnosed, we see many of

**Figure 12.1**   The House of Usher, by Arthur Rackham. The narrator is on horseback in the foreground.

the same symptoms that the Usher's report, such as general malaise, lack of interest in food and consequent weight loss, poor balance, visual disturbances (such as photophobia or sensitivity to light), depressed mood, itching skin and a changed sense of smell.[2-7]

## 12.2   THE POISON BEHIND THE STORY

Carbon monoxide is known as the "silent killer" because it is a gas with no taste, colour or smell. It also doesn't irritate the eyes or skin, like some other gases or nerve agents.[8] Carbon monoxide is produced naturally high up in the atmosphere,[9] but closer to earth is given off when fossil fuels such as oil, gas, coal, wood, and petrol are burned in confined and stuffy areas. When these fuels burn fully and cleanly, with enough oxygen, the greenhouse gas carbon dioxide ($CO_2$) comes off.[10] But if the fuel is burning in a space that lacks oxygen, or if the fire is smouldering, it doesn't produce enough heat and CO comes off instead. This is because oxygen is needed to drive the reaction all the way to $CO_2$ and without enough of it the carbon in the fossil fuel can only form CO. The CO-producing culprits in the Usher household are most likely leaky gaslights or gas lamps. As we have seen, these were filled with coal gas, which was also known as "illuminating gas" or "town gas" in the 19th Century. This is different to our modern domestic "natural gas" supply, which comes directly from the ground and is mostly methane ($CH_4$). Coal gas was made by energy companies by burning large amounts of coal in a confined space without oxygen, which meant it had a high concentration (8–16%[11]) of CO.[12] The coal gas was then carried into homes through a network of (often leaking) pipes. Coal gas was also used to heat early gas ovens, which were used for suicide. Before the dangers of illuminating gas were understood, it was popular in gas lighting because it burned with a such a bright blue flame (Figure 12.2).[10]

Coal gas was gradually replaced by natural gas in the 1960s and 1970s in the UK, making homes much safer. Poisonings with CO have declined since then, but continue to happen, with 108 deaths due to CO being recorded in England & Wales in 2021.[13] Deaths due to CO in the UK are now mostly from smoke inhalation during fires, or suicides involving car exhausts.[14] Thankfully, faulty gas boilers are now picked up at annual gas safety inspections, and household CO alarms, which although already fairly common, became mandatory in rented properties in UK in October 2022.[15]

**Figure 12.2**    A blue gas flame. Household gas is now natural gas ($CH_4$) but
                during the 1800s it would have been coal gas containing CO.

So why is CO so deadly? The answer lies in our red blood cells, which contain something called "haemoglobin". In a healthy person, oxygen ($O_2$) from the air combines with the haemoglobin and is carried around the body by the blood to our tissues and organs. When we add CO to the mix, it does a very convincing impersonation of oxygen, booting the $O_2$ off the haemoglobin and travelling through the bloodstream in its place. The bond between the CO and haemoglobin is so strong that this effect is hard to reverse. If the victim carries on breathing in CO, it is not long before the amount of $O_2$ in their body drops to dangerously low levels, causing them to suffocate. CO also binds to some important proteins in the heart, interfering with the heart's normal function.[16]

When presented with a suspected CO poisoning, rather than measure $O_2$ levels, toxicologists look at the complex formed between the iron in haemoglobin and CO, called "carboxyhemoglobin" or "COHb". Specifically, we measure the percentage of the haemoglobin that has become saturated with CO (%COHb). As a rule-of-thumb, the higher the value of the %COHb, the more severe the poisoning.

We all have low levels of COHb in our blood from breathing in traffic fumes, and because CO is a by-product of some normal biochemistry that happens in the body. These levels are usually <5% (so 5% of our haemoglobin has CO on it rather than $O_2$) but can be higher in cigarette smokers.[17] If we breathe in CO, the percentage creeps up and symptoms such as headache (10–20%), breathlessness, nausea and fatigue (20–30%), weakness (30–40%) and confusion (40–50%) start to appear.[18,19] A %COHb value of more than 50% is considered life-threatening because of the risk of respiratory collapse and coma. At a %COHb level of 50%, death can take less than 2 h without medical treatment.[18] If the poisoning is fatal, the skin of the deceased can sometimes be bright red,[5] but only in about a third of victims.[20] Interestingly, if the %COHb rises rapidly in an acute poisoning, the symptoms are less severe than a gradual rise to the same level.[21] There is not always a good correlation between %COHb and symptoms, and that is because the severity of poisoning depends on factors such as how long the victim has been exposed to the CO, and whether they have pre-existing cardiovascular problems (see Chapter 7).[14]

Typically, forensic toxicologists deal with acute CO poisoning cases from a sudden exposure to a high concentration of CO from a fire or car exhaust. On the other hand, chronic CO poisoning is a lengthy exposure to low levels of CO.[6] There is little agreement about what %COHb qualifies as chronic CO poisoning, with some studies using 5–10% as a guide. Chronic CO poisoning is poorly studied, and because the symptoms are so generic, they can often be mistaken for food poisoning, a virus, a hangover, drunkenness, drug overdose, epilepsy, or in older patients, dementia.[2,3] The blood test for COHb is quick and cheap (see Section 12.5) but is not always requested by hospital staff (in one study of CO poisonings it was requested in only a quarter of cases[22]). Because of this, it is possible that low-level exposure to CO is responsible for the symptoms of many unwell patients, but is being missed. We call this "occult" CO poisoning.[23]

Treatment for either type of CO poisoning involves removing the patient from the source of the poison and then fitting a mask with 100% oxygen flowing through it.[24] This gradually replaces the CO on the haemoglobin with $O_2$ but can take several hours. In more severe poisonings the patient is placed in a hyperbaric (high-pressure oxygen) chamber for around 90 min.[6] This works by getting more oxygen into the blood than would be possible by merely breathing it in at normal air pressures.[17]

### 12.3  POISONOUS PLOTS

Nowadays CO tends to be associated with suicides or house fires; we don't think of it as a murder weapon, but it was in the 19th and 20th Centuries when some murder victims were tied up in closed rooms with burning charcoal, or a leaking coal gas device.[12] The murderer then attempted to make the scene look like an accident.[10] Burning charcoal was a favourite plot line for murder mystery writers in this period, but it has re-emerged in the 21st Century as a real-life suicide method (see Section 12.5).

Shakespeare's *Julius Caesar* (1599) is most famous for the dictator's betrayal and death by stabbing, but there is also a fire-related death in the play. Brutus is Caesar's right-hand man (at least at the start of the play) but his wife Portia takes her own life before the coup is complete

BRUTUS: Impatient of my absence,
And grief that young Octavius with Mark Antony
Have made themselves so strong – for with her death
That tidings came. With this she fell distract
And, her attendants absent, swallowed fire.

Act IV, scene iii
*The Tragedy of Julius Caesar* by William Shakespeare,[25] 1599

The phrase "swallowed fire" might conjure up visions of a circus danger act, but there are two theories about what Shakespeare meant by this. One is that Portia swallowed hot coals then kept her mouth closed, with death from suffocation due to damage to her airway. The other, more likely, explanation is that she burned charcoal in a closed room and breathed in the fumes, dying of CO poisoning,[26] a known suicide method in Roman times.[27] For example, the Emperor Jovian died in 364 CE and one of the theories was that he died of CO poisoning from burning a large amount of charcoal.[28]

We see a vehicle exhaust suicide in the novel *Appointment in Samarra* (1934) by John O'Hara, about the self-destruction of a wealthy car dealer called Julian English. His is the first CO suicide to be reported in his fictional town of Gibbsville in the USA. The Deputy Coroner describes how

Mr English had had difficulties with Mrs English, so he went home and got drunk and while temporarily deranged through alcohol and grief, he, being well acquainted with the effects of carbon monoxide, being in the automobile business, why he committed suicide.

Chapter 10, p. 207
*Appointment in Samarra* by John O'Hara,[29] 1934

When Mr Julian English is found slumped in his running car in the garage by a neighbour, his face is described in the book as having a pinkish tinge. We have already seen that pink skin is indeed seen in CO fatalities, and this is thought to be caused by the red colour of $COHb$.[30] The neighbour who finds Mr English, reports a strong smell of whiskey on his breath. It's tempting to think that combining CO with alcohol would make matters worse, but the relationship between the two is quite complicated. Being under the influence of alcohol certainly clouds judgement and may prevent escape from a dangerous situation,[31] but other studies have found alcohol to play a protective role in CO poisoning. In one study of CO poisonings in the UK where alcohol had also been drunk, the %COHb reached at death was higher than for sober victims.[32] It's possible that drinking alcohol slows down the person's breathing meaning that they take in less CO.

The doctor who is called to the scene in *Appointment in Samarra* to certify the death, is the deceased's father Dr William English (chief of staff at the local hospital). In a modern-day death investigation a relative of the deceased would be very unlikely to be asked to attend the scene, but it could still happen in a small or remote community. In some countries there is only one forensic toxicology lab and scientists have to take care to avoid handling cases involving friends or relatives.

Zadie Smith's novel *White Teeth* (2000) opens with an unsuccessful suicide attempt using a car exhaust

At about 06.27 hours on 1 January 1975, Alfred Archibald Jones was dressed in corduroy and sat in a fume-filled Cavalier Musketeer Estate face down on the steering wheel,

Chapter 1, p. 3
*White Teeth* by Zadie Smith,[33] 2000

Archie is 'rescued' by the owner of the land he has parked his car on – the man needs the space for a meat delivery – who opens the car windows letting out the exhaust fumes.

Two of the deaths in *The Virgin Suicides* (1993) by Jeffrey Eugenides,[34] set in 1970s USA, involve CO. The novel is about five teenage sisters who all take their own lives. Mary puts her head in the oven, inhaling coal gas, and Lux inhales the exhaust fumes from the car in the garage. Interestingly, Lux's death is labelled as a CO poisoning by the author, but Mary's isn't. It seems that the author understood 'gassing yourself with the oven' to be a suicide method, but not that the mechanism of death was the same as using a car exhaust pipe.

When all of these novels were set, car engines were inefficient and the exhaust fumes contained up to 25% CO.[10] Exhaust deaths have declined since the widespread introduction of catalytic converters in the 1990s in the UK.[35] These have made exhaust fumes less dangerous by making sure the petrol burns fully into $CO_2$,[36] reducing the concentration of CO to 0.25% in modern car exhaust fumes.[37] Occasionally toxicologists will see a suicide attempt using a modern car exhaust. These are either unsuccessful and the person eventually turns to a different suicide method. However, sometimes the deceased dies due to $CO_2$ inhalation and asphyxiation instead.[38]

## 12.4 MODERN MEDICAL USES

It might seem unlikely that something so deadly could also have a medical use, but there are ongoing clinical trials investigating the potential of CO at the moment.[39] We know that CO is involved in the process of creating new memories in the brain.[9,37] It may also be that like nitric oxide (NO), CO could reduce inflammation, increase blood flow through arteries or dampen down the immune system in those receiving organ transplants.[37]

## 12.5 MODERN TOXICOLOGY CASES

Modern day accidental domestic CO home poisonings usually involve faulty gas boilers, generators, cookers and vehicles or other appliances (such as power tools) running inside closed garages.[40] Between 2015 and 2019, there were 2970 such non-fatal poisonings reported to the National Poisons Information Service in the UK.[41] Between 1998 and 2019 there were 750 fatalities related to accidental CO poisoning in England and Wales, the most common source being gas appliances.[42] More rarely, some workplace chemicals can lead to accidental CO poisoning such as the paint stripper dichloromethane, which is broken down in the body to CO,[5] and certain acids, which when mixed together form CO.[43,44] There are also some natural sources of CO such as forest fires,[45] thermal spas, volcanic eruptions and lightning strikes.[10] Another source of CO can be hookah or hubbly bubbly pipes, which burn charcoal.[21,46]

The most common source of CO poisoning however, remains fires, which can be accidental or deliberate.[14] In a series of 234 CO-positive deaths from Portugal between 2012 and 2014, the most common cause was domestic fire. Similarly in 59 CO-poisoning cases from Kuwait between 2014 and 2018, the majority (61%) were house fires.[47] The %COHb level can vary widely in these cases, which are also complicated by other factors such as serious burns and hydrogen cyanide inhalation (see Chapter 10), but the range in the Kuwaiti study was 50–80% (where >50% is life-threatening without medical treatment).[47]

Although suicides by exhaust fume inhalation have dropped because of various safety measures,[48] unfortunately other methods have taken its place. As we saw earlier, burning charcoal indoors can produce high concentrations of CO. The charcoal briquettes found in disposable BBQs or Japanese *hibachi* grills are particularly dangerous, as they are designed to smoulder rather than ignite.[49] If these are used as a suicide method and burned in a confined space, such as a car,[50] then higher %COHb levels are seen than for other sources of CO.[14] There can also be accidental charcoal CO poisonings, if the grill is used for cooking inside tents or camper vans. Hibachi suicides became more common in East Asian countries such as Japan, Hong Kong (where they were first reported[51]) and Taiwan in the mid-2000s, where it was widely publicised as a suicide method. Since then it has spread to Western countries, including the UK.[52]

Another suicide method known as "self-immolation" involves setting yourself alight with petrol, and this can also lead to significant %COHb levels. In a series of 32 such cases in Ontario, Canada, 7 had %COHb levels >50%.[53] The aptly named "death machine" is another CO-poisoning suicide method, and involves mixing together two acids and breathing in the fumes. In 2005 in the USA, a 21-year-old man was found dead inside a taped-up room with two buckets of chemicals. He was bright red and his %COHb level was later measured as 64%. The two chemicals were formic and sulphuric acid and had been bought on the internet.[54]

Homicides involving CO are not unheard of. In a series of four homicides in Japan, coal gas was used in two cases to poison unwanted lovers, and exhaust fumes in another. In the latter case, the body of the deceased was placed in a bathtub to give the impression he had drowned. As his %COHb level was 69% and he had no water in his lungs, this wasn't very convincing to the police. In the final case, a man killed his 43-year-old wife with CO in order to claim on her life insurance. The plan was highly elaborate and he had read two textbooks in an attempt to commit the perfect murder. He adopted a belt-and-braces approach, filling his greenhouse with both burning charcoal and the two acids mentioned above, and even testing the method on a rat beforehand. The police at first thought the death was accidental, but became suspicious after the life insurance payout, and the husband was sentenced to life in prison.[55] In a case in China in 2014, two men were found deceased inside a running car, they were bright red and both had %COHb levels of 70%. Initially the deaths were thought to be accidental, but later it was found that a large amount of cash the two deceased were carrying (for criminal reasons) had gone missing. They had met up with two people the night before to plan some criminal activity, but their coffee was spiked with estazolam (a sedative) and the money stolen. Once they had passed out, the two men were placed in the running car to inhale CO.[56]

These cases demonstrate that it is easy for investigators to jump to conclusions about the manner of death when they arrive at a scene. The narrative created at this stage can have far-reaching consequences for a case, especially if this version of events is passed to the forensic toxicologist. We know this phenomenon as "confirmation bias", and it is the natural tendency for humans to look for evidence that fits with what they already believe to be true, and to ignore evidence that doesn't fit. In the murders from China discussed above, a test for drugs was initially not deemed necessary because the cases were assumed to be CO accidents. Only later, when other evidence came to light, did the toxicologists return to the samples and test them further. Of course, it's not always the toxicologist who chooses the tests; sometimes the Pathologist or Coroner will send a list, which can be very limited.[57] Even if suspicions are raised, the case notes given to forensic toxicologists are collected at the very start of an investigation, and when new information comes to light, it is not always passed on. Further testing can be done later, but as we saw in Chapter 5, some drugs and poisons break down in body fluids in storage. Depending on the volume of the samples taken at the autopsy, there may not be enough left for another set of tests.[58]

## 12.6   CATCHING THE CHEMICAL CULPRIT

Carbon monoxide is not tested for in every toxicology case, but is targeted if the case circumstances suggest a fire, vehicle exhaust or the presence of burning charcoal. In some cases the Pathologist picks up that CO is potentially the culprit as the deceased's blood, tissues and organs are bright red.[10,59] Although the toxicologist does not attend the autopsy (see Chapter 7), clues like this, the presence of plant material in the stomach (Chapter 3), or the smell of almonds (Chapter 10) can help to shed light on what may have caused a death. We have met many different samples in this book including blood, urine, stomach contents and hair, but there are other sources of toxicology samples in the body that can be used if needed. For CO poisoning cases this can include the bright red organs described above. For example, a section of spleen may be taken, which is very high in red blood cells and so is a good place to look for COHb. Organs are not routinely used in forensic toxicology because they need so much clean-up, but it is not uncommon for a lab to receive a liver sample, particularly if the deceased is severely decomposed.[60] After death, body fluids tend to merge together, forming what is described as "cavity fluid". This can be analysed, but is very difficult to interpret, so a liver sample can be a better option. A small section is taken during the autopsy and homogenised in a blender by the toxicology lab. There are other case types where organs are useful, for example if there is little or no blood left after a traumatic accident or the victim cut their wrists or throat (known as "exsanguination"), or was badly burned in a fire.

One complication with testing samples for COHb is that once out of the body, the CO will detach from the haemoglobin and slowly rise to the top of the liquid sample and into the headspace of the tube (see Chapter 10). This is similar to your bottle of fizzy drink going flat in the fridge. Whenever you take the lid off, the tell-tale hissing noise alerts you to the escape of the $CO_2$. When this happens repeatedly to a blood sample containing CO, it can lead to falsely low COHb levels.[61] Some toxicology labs will try to prevent this by making sure the test for COHb is the first one they do.

The test for COHb is one of the simplest we have seen in this book, and is based on **spectrophotometry**. Some of the blood sample is placed into a glass tube and light of specific wavelengths is passed through it. The light will be absorbed by the COHb and any normal haemoglobin (called $O_2$Hb) that remains. The amount of light absorbed by each type of Hb tells the instrument what the ratio is between them and hence the %COHb saturation. This technique is called CO-oximetry and is the main one used by toxicology laboratories and hospitals,[8] although there are others such as **GC-MS**.[62] There are some emergency treatments that interfere with the CO-oximeter. For example, hydroxocobalamin (vitamin B12$_a$), which is given to fire victims suffering from cyanide poisoning (see Chapter 10) causes falsely low COHb readings.[17]

In an interesting case from the USA in 2020, three people were misdiagnosed with CO poisoning after breathing in some unknown fumes at work. After their (unnecessary) oxygen treatment and release from hospital, the CO-oximeter was found to be faulty and when their blood samples were re-analysed they were found to have normal %COHb levels.[63] Forensic scientists have a standard way of picking up faulty machines before any case samples are run, and so making this kind of error in a criminal case is much less likely. At the start of the testing, two "quality control" samples are run through the machine. One is a pre-made sample containing the drug or poison you are looking for. In the case of CO it would be a blood sample through which a certain amount of CO has been bubbled. We know what %COHb it has, so we check to make sure the machine gives the correct reading. This is known as a "positive" control, and reassures the scientist that their method is working correctly. The other sample is a "negative" control or "blank". This would be a blood sample containing no drugs or poisons and so shows you that your method has not been contaminated. If the sample comes up positive, you know there is an issue somewhere in the method.

Carbon monoxide can be detected in the air as well as the blood. If you have a CO detector in your house, it most likely contains a silica pad housing palladium chloride ($PdCl_2$). When CO wafts over it, it darkens as the palladium ions are reduced to palladium metal. The change in the amount of light entering the detector triggers the alarm.[9]

## 12.7 CASE CLOSED

In some of our previous case studies, the symptoms were either fleeting or missing entirely from the story. In *The Fall of the House of Usher*, the long and drawn-out illnesses the family suffer make the diagnosis easier. As we have seen, many of the symptoms Poe describes are similar to chronic CO poisoning, including photophobia (being "tortured by even a faint light"), feeling depressed and anxious, and having disorganised thought processes.[4] The pale skin colour of the Ushers has also been reported in previous CO poisonings.[64] The accuracy of the symptoms described may be because Poe himself was thought to be a long-term sufferer of chronic CO poisoning, likely due to the same leaky gas lamps.[39]

The story ends with Roderick Usher being frightened to death (probably suffering a heart attack) after seeing the 'ghost' of his sister Madeline. She had died and been placed in a coffin by him only a few days' earlier. But even this could be explained by chronic CO poisoning, as it has been thought to inspire ghostly visitations in the minds of its delirious victims.[9]

## REFERENCES

1. E. A. Poe, *The Fall of the House of Usher and Other Tales*, New American Library, New York, 1960.
2. C. J. Webb II and P. V. Vaitkevicius, *J. Am. Geriatr. Soc.*, 1997, **45**, 1281–1282.
3. M. V. Balzan, G. Agius and A. Galea Debono, *Postgrad. Med. J.*, 1996, **72**, 470–473.
4. R. A. M. Myers, A. DeFazio and M. P. Kelly, *J. Clin. Psychol.*, 1998, **54**, 555–567.
5. *Clarke's Analysis of Drugs and Poisons*, ed. A. C. Moffatt, D. Osselton and B. Widdop, Pharmaceutical Press, London, 4th edn, 2011.
6. L. A. Ruth-Sahd, K. Zulkosky and M. E. Fetter, *Dimens. Crit. Care Nurs.*, 2011, **30**, 303–314.
7. R. W. Byard, *Forensic Sci. Med. Pathol.*, 2019, **15**, 1–2.
8. D. E. Brooks, M. Levine, A. D. O'Connor, R. N. E. French and S. C. Curry, *Chest*, 2011, **140**, 1072–1085.
9. D. McMillan, in *Chemistry in its Element*, RSC Publishing, Cambridge, 2012.

10. D. Blum, *The Poisoner's Handbook: Murder and the Birth of Forensic Medicine in Jazz Age New York*, Penguin Books, London, 2010.

11. R. V. Clarke and P. Mayhew, *J. Crim. Justice*, 1988, **10**, 79–116.

12. J. F. O'Brien, *The Scientific Sherlock Holmes: Cracking the Case with Science & Forensics*, Oxford University Press, New York, 2013.

13. Deaths registered in England and Wales: 2021, https://www.ons.gov.uk/peoplepopulationandcommunity/birthsdeathsandmarriages/deaths/bulletins/deathsregistrationsummarytables/2021, accessed August 2022.

14. C. Forés Lisbona and H. J. Hamnett, *J. Forensic Sci.*, 2018, **63**, 1776–1782.

15. HM Government, *The Smoke and Carbon Monoxide Alarm (Amendment) Regulations 2022*, Home Office, London, 2022.

16. D. K. Quinn, S. M. McGahee, L. C. Politte, G. N. Duncan, C. Cusin, C. J. Hopwood and T. A. Stern, *J. Clin. Psychiatry*, 2009, **11**, 74–79.

17. L. S. Nelson, M. A. Howland, N. A. Lewin, S. W. Smith, L. R. Goldfrank and R. S. Hoffman, *Goldfrank's Toxicologic Emergencies*, McGraw Hill, New York, 11th edn, 2018.

18. M. Goldstein, *J. Emerg. Nurs.*, 2008, **34**, 538–542.

19. T. Struttmann, A. Scheerer, T. S. Prince and L. A. Goldstein, *J. Am. Board Fam. Pract.*, 1998, **11**, 481–484.

20. M. Costa, B. S. Silva, F. C. Real and H. M. Teixeira, *Forensic Sci. Int.*, 2019, **299**, 1–5.

21. *Disposition of Toxic Drugs and Chemicals in Man*, ed. R. C. Baselt, Biomedical Publications, Seal Beach, 12th edn, 2020.

22. National Poisons Information Service, *National Poisons Information Service Report 2020/21*, Public Health England, Edinburgh, 2021.

23. J. Wright, *Emerg. Med. J.*, 2002, **19**, 386–390.

24. T. Hargreaves, *Poisons and Poisonings: Death by Stealth*, RSC Publishing, Cambridge, 2017.

25. W. Shakespeare, *Julius Caesar*, Cambridge University Press, Cambridge, 2004.

26. K. Harkup, *Death by Shakespeare: Snakebites, Stabbings and Broken Hearts*, Bloomsbury, London, 2020.

27. S. W. Dubrey, O. Chehab and S. Ghonim, *Br. J. Hosp. Med.*, 2015, **76**, 159–162.

28. J. W. Drijvers, in *The Forgotten Reign of the Emperor Jovian (363–364): History and Fiction*, ed. J. W. Drijvers, Oxford University Press, Oxford, 2022, ch. 5, p. 109.

29. J. O'Hara, *Appointment in Samarra*, Penguin Books, New York, 1934.

30. L. D. Prockop and R. I. Chichkova, *J. Neurol. Sci.*, 2007, **262**, 122–130.

31. P. Sharma and D. G. Penney, *Toxicology*, 1990, **62**, 213–226.

32. L. A. King, *Hum. Exp. Toxicol.*, 1983, **2**, 155–157.

33. Z. Smith, *White Teeth*, Penguin, London, 2000.

34. J. Eugenides, *The Virgin Suicides*, Farrar Straus Giroux, New York, 1993.

35. G. M. G. McClure, *Br. J. Psychiatry*, 2000, **176**, 64–67.

36. T. Stone and G. Darlington, *Pills, Potions, Poisons: How Drugs Work*, Oxford University Press, Oxford, 2000.

37. J. Emsley, *Molecules of Murder: Criminal Molecules and Classic Cases*, RSC Publishing, Cambridge, 2008.

38. G. A. Schmunk and J. A. Kaplan, *Am. J. Forensic Med. Pathol.*, 2002, **23**, 123–126.

39. L. E. Otterbein, *Med. Gas Res.*, 2013, **3**, 7.

40. S. S. Zimmerman and B. Truxal, *Pediatrics*, 1981, **68**, 215–224.

41. D. Gentile, R. Adams, M. Klatka, S. Bradberry, L. Gray, R. Thanacoody, G. Jackson and E. A. Sandilands, *J. Public Health*, 2022, **44**, 565–574.

42. R. M. Close, N. Iqbal, S. J. Jones, A. Kibble, R. J. Flanagan, H. Crabbe and G. S. Leonardi, *Int. J. Environ. Res. Public Health*, 2022, **19**, 4099.

43. D. Fishwick, M. Carder, I. Iskandar, B. C. Fishwick and M. van Tongeren, *Occup. Environ. Med.*, 2022, **79**, 628–630.

44. J. P. Ferrando, A. F. Dufol, S. Nogué, F. Cordoba, C. Merino and A. Supervía, *40th International Congress of the European Association of Poisons Centres and Clinical Toxicologists (EAPCCT)*, Tallinn, Estonia, 2020, p. 584.

45. W. P. Putra and A. Pribadi, *IOP Conf. Ser.: Earth Environ. Sci.*, 2021, **871**, 012027.

46. S. S. Retzky, *J. Med. Toxicol.*, 2017, **13**, 193–194.

47. A. Al-Matrouk, A. Al-Hemoud, M. Al-Hasan, Y. Alabouh, A. Dashti and H. Bojbarah, *Int. J. Environ. Res. Public Health*, 2021, **18**, 8854.

48. V. Routley, *Crisis*, 2007, **28**, 28–35.

49. E. F. Wilson, T. H. Rich and H. C. Messman, *J. Am. Med. Assoc.*, 1972, **221**, 405–406.

50. S. O'Donovan, C. van den Heuvel, M. Baldock and R. W. Byard, *Med. Sci. Law*, 2022, DOI: 10.1177/00258024221122187.

51. C. M. Leung, W. S. Chung and E. P. So, *J. Clin. Psychiatry*, 2002, **63**, 447–450.

52. Y.-Y. Chen, O. Bennewith, K. Hawton, S. Simkin, J. Cooper, N. Kapur and D. Gunnell, *J. Public Health*, 2013, **35**, 223–227.

53. M. J. Shkrum and K. A. Johnston, *J. Forensic Sci.*, 1992, **37**, 208–221.

54. J. A. Prahlow and B. W. Doyle, *Am. J. Forensic Med. Pathol.*, 2005, **26**, 177–180.

55. S. Akaishi, S. Oshida, K. Hiraiwa, I. M. Sebetan, Y. Ohno, F. Kuroda, T. Suzuki and S. Kashimura, *Z. Rechtsmed.*, 1982, **88**, 297–304.

56. N. Zhongyu, Z. Mengjun and D. Zhenhua, *Forensic Sci. Technol.*, 2020, **45**, 325–327.

57. *Principles of Forensic Toxicology*, ed. B. Levine, AAAC Press, Washington DC, 2013.

58. H. J. Hamnett and I. E. Dror, *Forensic Sci. Int.: Synergy*, 2020, **2**, 339–348.

59. M. Janík, M. Ublová, Š. Kučerová and P. Hejna, *J. Forensic Leg. Med.*, 2017, **48**, 23–29.

60. R. J. Dinis-Oliveira, F. Carvalho, J. A. Duarte, F. Remião, A. Marques, A. Santos and T. Magalhães, *Toxicol. Mech. Methods*, 2010, **20**, 363–414.

61. D. H. Chace, L. R. Goldbaum and N. T. Lappas, *J. Anal. Toxicol.*, 1986, **10**, 181–189.

62. S. Oliverio and V. Varlet, *J. Anal. Toxicol.*, 2018, **43**, 79–87.

63. S. L. Thornton, W. Barkman and T. Rianprakaisang, *American College of Medical Toxicology 2020 Annual Scientific Meeting New York*, 2020, p. 131.

64. M. F. Hanif, B. Iqbal and N. Gilani, *J. Pak. Med. Assoc.*, 2016, **66**, 771–773.

# Subject Index

Locators in *italic* refer to figures; those in **bold** to glossary
TCM = traditional Chinese medicine